Hello, World!

Hello, World!

A Brief History of Programming in 90 Languages

Dale Biagio

Badger Media LLC

Hello, World! A Brief History of Programming in 90 Languages

by Dale Biagio

Edited by Natasha Biagio

First edition, 2026

All code samples in this book reproduce the minimal "Hello, World!" program for each language as documented in official language specifications, tutorials, or reference implementations. Code samples are presented for educational and historical purposes.

Language creation dates reflect the year of first public release or publication. Where dates are disputed, the earliest credible source is used.

Every code sample in this book has been verified against the language's official documentation or reference implementation.

Historical accounts, biographical details, and quotations are drawn from published interviews, official documentation, academic papers, news reporting, and widely cited industry sources. Statistics reflect widely reported industry estimates.

Typeset in Lora, Liberation Sans, and Noto Sans Mono.
Printed and bound by print-on-demand.

ISBN 979-8-9950721-0-2 (paperback)
Library of Congress Control Number: 2026909063

helloworldthebook.com

Advance Praise for Hello, World!

"*As far as I am concerned it is perfect.*"

— Martin Richards, creator of BCPL

"*A remarkably readable romp through the development of programming languages. While languages are the critical tool of the most indispensable profession of our era, this is not a boring history of technical arcana. Rather, Biagio gives us a revealing glimpse into the rooms where it happened, showing how paradigms and products evolved from the needs of practitioners and brilliant inventors. It's packed with vignettes of creators, their motivations, their achievements, and their failures. Like me, I know you will learn more than you expected, and have fun doing it.*"

— Alan Cooper, creator of Visual Basic

"*Biagio captures the human stories behind 90 programming languages with genuine care and accuracy. A valuable contribution to the history of computing.*"

— Don Chamberlin, co-creator of SQL

"*Many of us are familiar with some programming languages, but few of us know ninety of them. Biagio takes us on an engaging, thoughtful, and sometimes quirky tour of a huge range of languages, enlivened with human stories about their origins and creators.*"

— Simon Peyton Jones, co-designer of Haskell

Contents

Era 3: The Expansion (1980–1989)

Era 4: The Internet Age (1990–1999)

Era 5: The Oddballs (varied)

Era 6: The Modern Era (2000–2009)

Era 7: The Safety Era (2010–2019)

Era 8: The AI Age (2020–)

A SPOTLIGHT ON B AND C

The Day Hello World Was Born

In 1972, Brian Kernighan sat down at Bell Labs to write an internal tutorial for the programming language B. He needed a simple example, something that would prove the toolchain worked without demanding any real knowledge of the language. He typed a short program that printed two words to the terminal: hello, world.

Kernighan did not think of this as a historic moment. The phrase was modest, lowercase, unpunctuated. He may have borrowed it from a cartoon he barely remembered, a chick hatching from an egg and saying "Hello, World." Years later, he couldn't recall the exact source. It was just a placeholder. A throwaway.

Six years later, Kernighan and Dennis Ritchie published The C Programming Language, a book that would sell millions of copies and define how an entire generation learned to program. Chapter 1 opened with the same example, almost unchanged:

```
#include <stdio.h>

main()
{
    printf("hello, world\n");
}
```

The original C "hello, world" from The C Programming Language (1978)

The book's influence was extraordinary. Universities adopted it immediately. Within a decade, nearly every programming textbook began with a Hello World example. The phrase migrated from C to Pascal, from Pascal to Java, from Java to Python. It became an industry term, "Time to Hello World" measures how quickly a developer can get a new language running.

Hello World became the first thing virtually every programmer writes. It has been printed to screens in hundreds of languages, on devices from supercomputers to microcontrollers. It is simultaneously the simplest program and the most universal one, a two-word greeting that has been repeated billions of times in every human language, on every continent, by beginners and experts alike.

All because Brian Kernighan needed a throwaway example and reached for something friendly.

Introduction

"Hello, World!" has been written in just about every programming language on the planet. It has been printed to screens on every continent, on devices from supercomputers to wristwatches. It is simultaneously the simplest program and the most universal one, a two-word greeting that has been repeated countless times since Brian Kernighan first wrote it in a Bell Labs tutorial in 1972.

This book tells the story of 90 languages that learned to say it.

From Konrad Zuse's Plankalkül, designed in wartime Germany and formalized in 1948 but never implemented in his lifetime, to languages released in 2024, each entry presents the code, the creator, the year, and the human story that gave the language its reason to exist. They are sorted chronologically within era chapters, so reading front to back is reading the history of how humans learned to instruct machines.

You can see paradigms rise and fall in the syntax. Structured programming arrives and GOTO disappears. Object orientation sweeps through, then functional programming stages a comeback. Each shift is visible in how languages say "Hello, World!", from FORTRAN's WRITE statement to Haskell's putStrLn to Rust's println! macro.

Every language in this book was created because someone believed there was a better way to talk to a machine.

This book is curated, not a census. Thousands of languages deserve entries; ninety earn theirs through two filters: impact and story. Some changed how millions of programmers think: C, Python, JavaScript. Others pioneered a paradigm or solved a problem so elegantly they influenced everything after: Lisp, Haskell, Go. Still others carry human narratives that teach us why languages exist at all: Plankalkül's lonely wartime genius, Rust's answer to C++'s vulnerabilities, Julia's reach toward scientific computing. Scan the list and you will notice absences. You are right. The end matter includes a page called "The Ones That Got Away," honoring languages that missed this volume but left their mark. There is no universal answer, which is why this book holds ninety languages instead of one.

Each language gets one page: code, creator, year, and a pocket history. For a few, that is not enough. The spotlights go deeper. They pause on a language and its world, exploring the debates that shaped it, the people who fought for it, the problems it solved in its time, and the way it echoes in code today. A spotlight on Smalltalk is not just the language but the vision of personal computing it embodied. Haskell's spotlight talks about lazy evaluation and the academic idealism that protected it while industry raced elsewhere. These essays ask why a creator made the choices they did, and what those choices reveal about the era in which the language was born.

This book moves between entry and essay. You read a one-page account of ALGOL, then turn the page to find a spotlight on the languages it spawned. You meet a four-line sample in C, then read ten pages about the tensions between portability and power that C embodied. The rhythm is intentional. The entries are the spine of history, in chronological order. The spotlights are the nervous system, adding depth where the story demands it. Neither alone is complete. The entries teach you the breadth; the spotlights teach you why breadth matters.

This is part reference, part narrative. Use it as a desk dictionary: open to Python, find the code, read the three-paragraph origin story, and move on. Or read it front to back and watch languages rise and fall, paradigm by paradigm. You will see what survives and what fades. You will see that some of the most influential languages came from one person with a clear vision, while others emerged from committee compromise. You will see that a language is never neutral. Every syntax choice, every design decision, carries the fingerprints of its makers and the pressures of its moment. This book is about seeing, in the code we write every day, the human choices that built it.

Further Reading

The History of Programming Languages (HOPL) conferences are the canonical academic record of the field. These ACM SIGPLAN conferences, held roughly every decade since 1978, gather creators and historians to reflect on how languages came to be. The proceedings are primary sources: papers written by the people who designed the languages, papers arguing about causation and influence, papers preserving the memory of languages that faded. They are essential reading for anyone who wants to understand why programming languages exist. This book draws most heavily on HOPL II and HOPL III. HOPL I is the historical foundation. HOPL IV, published open-access, is the most recent gathering.

HOPL I (1978). History of Programming Languages. Edited by Richard L. Wexelblat. Proceedings of the ACM SIGPLAN History of Programming Languages Conference, Los Angeles, CA, June 1 to 3, 1978. Published by Academic Press / ACM, 1981 (ACM Monograph Series). ISBN 0-12-745040-8.

HOPL II (1993). The Second ACM SIGPLAN Conference on History of Programming Languages. Conference Chair John A. N. Lee. Program Chair Jean E. Sammet. Cambridge, MA, April 20 to 23, 1993. ACM Proceedings DOI 10.1145/154766. https://dl.acm.org/doi/proceedings/10.1145/154766. ISBN 978-0-89791-570-0. Book edition: History of Programming Languages, edited by Thomas J. Bergin and Richard G. Gibson. Addison-Wesley, 1996.

HOPL III (2007). Proceedings of the Third ACM SIGPLAN Conference on History of Programming Languages. Conference Co-Chairs Barbara Ryder and Brent Hailpern. San Diego, CA, June 9 to 10, 2007. ACM Proceedings DOI 10.1145/1238844. https://dl.acm.org/doi/proceedings/10.1145/1238844. ISBN 978-1-59593-766-7.

HOPL IV (2021). Proceedings of the ACM on Programming Languages, Volume 4, Issue HOPL, June 2020. Co-Chairs Guy L. Steele Jr. and Richard P. Gabriel. Conference held virtually June 20 to 22, 2021 (postponed from 2020 due to COVID-19). Open Access. https://dl.acm.org/toc/pacmpl/2020/4/HOPL.

Note on Sources

A full bibliography appears at the back of this book, organized by language. It includes primary sources (academic papers, language specifications, the oral histories recorded at the History of Programming Languages conferences), secondary histories written by designers and historians, online archives and documentation repositories, and the conference proceedings themselves. Use it as a jumping-off point. Every citation is an invitation to read deeper. Some papers sit behind paywalls; many are free through university libraries, archive sites, or the authors' own pages. The field of programming languages is built on conversation and documentation. The bibliography is your map into that conversation.

How to Read This Book

Each of the 90 languages in this book gets one page.

1 **LANGUAGE NAME** YEAR

2 Creator Name ● Active

3 The human story: who created this language, why they were frustrated with what existed, and what it changed. Two to four sentences that make you care.

4
```
#include <stdio.h>
int main() { printf("Hello, World!\n"); }
```

5 *The #include imports the standard I/O library; printf sends text to the screen.*

6 **Paradigm:** Procedural, imperative

Typing: Static, weak

Usage: ●●●●●

Lines: 5

7 ★ Dennis Ritchie and Ken Thompson created C while building Unix, each tool shaped the other.

---

1 **Language name** : Large, bold, top left; year of first release, top right

2 **Creator and status** : Filled dot = Active, half = Endangered, empty = Extinct

3 **Description** : The human story: who, why, and what it changed

4 **Code block** : The complete Hello World program in monospace

5 **Code note** : Optional plain-English explanation of non-obvious syntax

6 **Metadata** : Paradigm, type system, and usage (5-dot scale)

7 **Fun fact** : One surprising, memorable, verifiable detail

Usage scale: 1 dot = historical/esoteric, 5 dots = ubiquitous (based on TIOBE, Stack Overflow, and GitHub data).

A SPOTLIGHT ON LISP

The Notation That Came Alive

John McCarthy, the Dartmouth Summer, and the Language That Was Never Meant to Run

When John McCarthy sat down to design a programming language, he was not thinking about programmers. He was thinking about mathematics. Specifically, he was thinking about Alonzo Church's lambda calculus, a formalism so theoretical, so stripped down to its essence, that it existed in the same realm as pure logic and set theory. McCarthy had no plans for LISP to actually run on a computer. It was supposed to be notation, the way mathematical equations are notation: elegant, precise, and safely confined to the page. The idea that anyone would actually implement it and use it to solve practical problems would have struck McCarthy as almost comically beside the point.

Yet in 1958, when McCarthy was at the Massachusetts Institute of Technology developing artificial intelligence research alongside Marvin Minsky and others, he began asking himself a deceptively simple question. If you wanted to build a machine that could think, what language would that machine need? Forget the language humans would use to program it; what conceptual language would the thinking itself require? McCarthy's answer was revolutionary not because it was complicated, but because it was radical in its minimalism. He would design a language built on just a handful of primitives, atoms, lists, conditional expressions, and recursion, that could, through their combination, express any computation at all.

This impulse toward reduction, toward finding the irreducible core of a thing, had deep roots in McCarthy's life. His parents, John Patrick McCarthy and Ida Glatt McCarthy, were both members of the Communist Party of America, steeped in a philosophy that sought to understand history and society through first principles. His father had become the business manager of the Daily Worker, the party's official newspaper. Growing up in this atmosphere of ideological intensity, McCarthy learned to ask fundamental questions, to challenge assumptions, to burrow down to the essential. He had briefly joined a communist cell as a graduate student, but withdrew, finding its rigid orthodoxies unable to accommodate his restless intellect. Later, after the Soviet invasion of Czechoslovakia in 1968, he would move definitively rightward, becoming a conservative Republican. But the imprint of his early years, that hunger to reduce complexity to first principle, never left him.

The path to LISP ran through the summer of 1956 at Dartmouth College. McCarthy, along with Marvin Minsky, Nathaniel Rochester, and Claude Shannon, had written a proposal for a summer research project on artificial intelligence. The term itself, artificial intelligence, was theirs. No one had called it that before. The conference was audacious in scope and spirit: they wanted to explore whether machines could be made to simulate human intelligence. McCarthy brought his mathematical background, his hunger for fundamental principles, and his growing conviction that machines could be programmed to use symbolic reasoning the way humans did.

By 1958, the direction was clear. McCarthy needed a language for symbolic computation, one that could operate on symbols and lists as easily as arithmetic languages operated on numbers. He borrowed notation from Church, the lambda calculus provided a way to express functions formally and compactly. But McCarthy did not attempt to implement Church's full theoretical apparatus. "I didn't understand the rest of the book," he later recalled, "so I wasn't tempted to try to implement his more general mechanism." Instead, McCarthy extracted what he needed: the lambda notation for functions, and the idea that functions could be treated as data, passed as arguments, returned as values.

What emerged was staggering in its elegance. McCarthy defined a handful of primitive operations: atom (to test if something is an atom), eq (to test equality), cons (to construct a list from an element and another list), car (to extract the first element of a list), and cdr (to extract the rest of a list). He added the conditional expression, if this, then that, else the other, and the ability to define and call function recursively. That was all. Everything else, McCarthy proved, could be built from these pieces. The language could compute anything that was computable. It was Turing-complete, though that term had its own curious history of not being quite understood until others had implemented it. McCarthy published his formal design in April 1960 in Communications of the ACM, in a paper titled "Recursive Functions of Symbolic Expressions and Their Computation by Machine, Part I." The paper was dense and mathematical, the language of the logician, not the engineer. McCarthy had created a notation, a theoretical artifact, the kind of thing you might study as a mathematician studies group theory. He had shown, rigorously, that you could express any recursive function using his few primitives combined with symbolic expressions. It was beautiful and impeccable and entirely theoretical. McCarthy expected it to remain theoretical.

This is where Steve Russell, one of McCarthy's students, entered the story. Russell read McCarthy's paper carefully. He studied the eval function, the universal evaluator that McCarthy had defined as a thought experiment, a way of proving that the language could interpret itself, could take its own code as data and compute with it. This was a purely theoretical flourish, an elegant demonstration of mathematical power, like a proof that certain numbers have certain properties.

And Russell, bright-eyed graduate student that he was, looked at the eval function and saw something no one else had imagined. He saw that you could actually build this. You could take McCarthy's mathematical notation, translate it into the machine code of an IBM 704, and create an actual interpreter that actually ran LISP programs.

McCarthy's reaction has become legend in computing history. Russell went to McCarthy and said he would code up the eval function. McCarthy reportedly replied with something like: "You're confusing theory with practice. This eval is intended for reading, not for computing." But Russell went ahead anyway. He hand-compiled the eval function into IBM 704 machine code, worked through the bugs, and created the first LISP interpreter. When McCarthy saw what Russell had done, he was startled. What had been notation suddenly became code. What had been theory became practice. A language that was never supposed to run had run. A mathematical formalism had escaped onto the machine and survived.

What happened next shaped the history of computing. LISP became the language of artificial intelligence research, the default idiom for anyone trying to make machines think. More than that, it introduced ideas that still dominate computer science today. Garbage collection, the idea that the machine itself could reclaim memory from data structures no longer in use, came directly from LISP. Higher-order functions, the ability to treat functions as values and pass them around, became a native feature of LISP and would later revolutionize other languages. Dynamic typing, the freedom from having to declare types in advance, was built into LISP from the start. Recursion as a primary tool for control flow rather than iteration, LISP made this natural and graceful. Tree data structures, linked lists, the read-eval-print loop where programmers type code and see results immediately: all of these emerged from LISP or were perfected by it.

McCarthy himself would continue to develop these ideas, moving to Stanford, where in 1965 he founded the Stanford Artificial Intelligence Laboratory. In 1971, at age forty-three, he received the Turing Award, the highest honor in computer science, for his contributions to artificial intelligence and the design of LISP.

But the deepest impact of LISP was not something McCarthy had planned or even fully anticipated. He had set out to create a notation, a theoretical tool.

What he created instead was a language that revealed something profound about the relationship between mathematics and physical machines. A notation born in mathematical abstraction, grounded in lambda calculus and recursive function theory, could be yanked into reality and made to run on physical hardware. The gap between pure theory and practical implementation, which McCarthy thought was absolute, turned out to be bridgeable. A student, looking at equations on a page, could see how to translate them into machine code. The notation came alive.

More than half a century later, descendants of LISP like Scheme and Clojure would still be expressing those same core ideas: atoms, lists, functions as first-class values, recursion, and the profound unity between code and data. McCarthy had designed a language for thinking, and it turned out to be exactly the language machines would learn to think in.

What happened next

Year	Event
1956	Dartmouth Summer Research Project on AI; McCarthy and colleagues coin the term "artificial intelligence"
1958	McCarthy begins designing LISP at MIT; sketches the language based on lambda calculus and S-expressions
1959	Steve Russell creates the first LISP interpreter by hand-compiling McCarthy's eval function to IBM 704 machine code
1960	McCarthy publishes "Recursive Functions of Symbolic Expressions and Their Computation by Machine, Part I" in Communications of the ACM
1965	McCarthy founds the Stanford Artificial Intelligence Laboratory (SAIL); becomes a full professor at Stanford
1968	Soviet invasion of Czechoslovakia influences McCarthy's political shift from progressive to conservative Republican
1971	McCarthy receives the ACM Turing Award for contributions to artificial intelligence
1980	LISP becomes the dominant language for AI research; LISP machines emerge as specialized computers optimized for symbolic processing
2011	John McCarthy dies at age 84; his legacy as founder of AI and designer of LISP remains central to computer science

Era 1

The Pioneers

1948–1959

The birth of programming languages marked humanity's first attempts to communicate with machines in something resembling human terms. Before high-level languages, programming meant manipulating switches and wiring circuits by hand, a process that required deep knowledge of the machine's internals. Konrad Zuse's Plankalkül represented the first theoretical framework for automatic computation, though his work remained hidden in the isolation of postwar Germany. Kathleen Booth and David Wheeler pioneered Assembly. Bill Atkinson's HyperTalk (1987) is included here for its conceptual link to these pioneers of human-machine interaction.

These languages were born of practical necessity, not grand vision. Early computers were rare, expensive, and temperamental machines that required entire rooms and consumed massive amounts of electricity. Programming them was an art practiced by a tiny priesthood of engineers who understood circuits and pulses. Fortran's success showed that high-level languages could match hand-written assembly for speed, shattering the myth that efficiency required sacrificing abstraction. Then the floodgates opened: LISP brought functional programming and the radical idea that code could be data; COBOL brought English-like syntax to business computing; RPG made report generation accessible to operators rather than programmers. By 1959, the era of direct hardware manipulation was ending, and the diversity of approaches (mathematical, business-oriented, practical) signaled that there would never be just one way to program a computer.

9 languages in this era

Timeline

- 1948 — Konrad Zuse proposes Plankalkül
- 1949 — Booth and Wheeler pioneer Assembly language
- 1955 — Grace Hopper creates FLOW-MATIC
- 1956 — Allen Newell, Cliff Shaw, Herbert Simon create IPL for AI research
- 1957 — Fortran 1.0 released by IBM; John Backus and team prove high-level languages can match assembly speed
- 1958 — LISP created by John McCarthy at MIT
- 1959 — COBOL created by Grace Hopper team; IBM releases RPG

World Context

Post-WWII computing boom; Korean War; Sputnik (1957); dawn of the space race

Dominant Paradigms

Imperative programming dominant; first functional language (LISP); English-like syntax for business (COBOL); machine-centric thinking giving way to human-readable code

A SPOTLIGHT ON FORTRAN

The Dropout and the Priesthood

John Backus, IBM, and the Language That Proved Machines Could Understand Math

In 1954, programming a computer meant hand-to-hand combat with the machine. You wrote instructions in machine code or assembly, raw numerical sequences, each one mapped to a specific operation on a specific piece of hardware. A single misplaced digit could crash the program. A simple mathematical formula might require dozens of instructions to encode. And if you wanted to move your program to a different machine, you started over.

The people who could do this work were scarce and they knew it. IBM estimated that the cost of programming had begun to rival the cost of the hardware itself. Programmers formed what John Backus later called a "priesthood", an elite guild that guarded its expertise and viewed any attempt to simplify the process with hostility and derision. They had spent years learning to speak the machine's language. They were not interested in teaching the machine to speak theirs.

Backus was an unlikely candidate to challenge them. Born in Philadelphia in 1924, he drifted through his early life with no particular direction. He entered the University of Virginia to study chemistry and was expelled for poor attendance. He enlisted in the Army during the war, was sent through a series of aptitude-driven reassignments, and wound up studying engineering, then medicine at Hahnemann Hospital in Philadelphia. There, doctors discovered a cranial bone tumor. They operated, replacing part of his skull with a metal plate. After the surgery, he abandoned medicine and enrolled at Columbia University to study mathematics, almost on a whim. In 1949, while walking through midtown Manhattan, he wandered into the IBM Computing Center on Madison Avenue, saw an electronic calculator on display, and struck up a conversation with a guide. By the end of the visit, he'd been given an aptitude test. He passed. IBM hired him as a programmer, and Backus, the aimless dropout with a metal plate in his head, found the thing he was good at.

He was very good at it. By 1953, Backus had already created Speedcoding, a system that made the IBM 701 easier to program but ran twenty times slower than hand-coded instructions. That trade-off, programmer convenience versus machine efficiency, was the central tension of the era, and every previous attempt to resolve it had failed. High-level systems were slow. Fast code was handwritten. The priesthood considered this an immutable law.

In late 1953, Backus proposed something audacious: a system that would let scientists write mathematical formulas in something close to algebraic notation, and have the machine translate those formulas into code as efficient as anything a human programmer could produce by hand. He called the project FORmula TRANslation, FORTRAN. IBM management was skeptical but gave him a small team and expected results within six months.

The team Backus assembled was deliberately eclectic. There were ten of them: a Vassar mathematics graduate named Lois Haibt, the only woman on the team; Roy Nutt, who would design the critical FORMAT statement for controlling output; Harlan Herrick, who would run the first Fortran program; a cryptographer; a chess champion; a crystallographer. Backus later recalled: "I figured there were a lot of smart people doing a lot of smart things, and we could learn from all of them."

Six months passed. The project was nowhere near done. The core challenge, building a compiler that could optimize code well enough to match human programmers, turned out to be far harder than anyone anticipated. The team used techniques borrowed from operations research, including Monte Carlo simulations, to analyze how programs actually executed and figure out where to focus optimization. It was pioneering work. No one had built anything like it before, because no one had believed it was possible.

Three years after the project began, in April 1957, the first Fortran compiler was delivered to a customer: Westinghouse-Bettis Atomic Power Laboratory, which used it for nuclear reactor design. The compiler produced code that was, in many cases, indistinguishable in speed from hand-optimized assembly. It held the optimization record for two decades. The priesthood's immutable law had been broken.

Adoption was swift. Within a year, more than half of all machine code written for the IBM 704 was being produced by the Fortran compiler rather than by hand. Scientists who had never programmed before could now write working code in an afternoon. The economics of software development shifted permanently, human time was expensive, machine time was getting cheaper, and Fortran proved you could stop wasting the former without sacrificing the latter.

Backus received the Turing Award in 1977. His acceptance lecture was characteristically contrarian. Instead of celebrating Fortran, he attacked the entire paradigm of imperative programming that he had helped establish, delivering a paper titled "Can Programming Be Liberated from the von Neumann Style?" He spent his remaining decades exploring functional programming, an intellectual repudiation of his most famous creation. The man who had freed programmers from machine code spent the rest of his career trying to free them from the way of thinking that machine code had imposed.

He died on March 17, 2007, at eighty-two. Fortran, his first and most consequential creation, was still running climate models, particle physics simulations, and weather forecasting systems, sixty-eight years old and still faster than most alternatives for the work it was designed to do. The dropout had built something the priesthood said was impossible, and it outlasted them all.

What happened next

Year	Event
1966	FORTRAN 66 standardized by ANSI, establishing consistent compiler behavior across vendors
1978	FORTRAN 77 released with character data types and block IF statements
1991	Fortran 90 adopted by ISO, introducing free-form source code, modules, and array operations
1997	Fortran 95 refines array handling and adds the FORALL statement for parallel-friendly loops
2004	Fortran 2003 standardized with object-oriented programming and C interoperability
2010	Fortran 2008 adds coarrays for parallel programming, reinforcing HPC dominance
2018	Fortran 2018 enhances parallel features for exascale computing
2023	Fortran 2023 published with conditional expressions, enumerations, and refined generics

PLANKALKÜL 1948

Konrad Zuse ○ Extinct

Konrad Zuse conceived Plankalkül (German for 'plan calculus') during WWII and formalized it in 1948, making it the first high-level programming language ever designed. It featured arrays, records, and conditional expressions decades before any other language. But postwar Germany was isolated, and the manuscript gathered dust until the 1970s, when researchers realized Zuse had invented programming language theory before anyone else.

```
V0 + V1 => V2
K1.2(V0) => R0
```

Zuse's variable notation; Plankalkül had no I/Oüit was pure computation

Paradigm: Imperative

Typing: Static

Usage: ●○○○○

Lines: 2

★ Plankalkül wasn't implemented until 2000, over fifty years after Zuse designed it. A team at the Free University of Berlin finally built a working compiler, confirming that his 1940s design actually worked.

ASSEMBLY 1947

Kathleen Booth, David Wheeler ● Active

Kathleen Booth wrote what is considered the first assembly language in 1947, working on the ARC computer at Birkbeck College, London, translating mathematical notation into machine instructions so humans could program without memorizing binary codes. She also wrote one of the first textbooks on programming. David Wheeler independently built the first practical assembler for EDSAC at Cambridge in 1949, creating the stored-program workflow that every computer still uses. Together, they established the thin layer between human thought and machine code, every language in this book compiles down to something like Assembly.

```
section .data
  msg db "Hello, World!", 10
section .text
  global _start
_start:
  mov rax, 1
  mov rdi, 1
  lea rsi, [msg]
  mov rdx, 14
  syscall
  mov rax, 60
  xor rdi, rdi
  syscall
```

x86-64; rax=syscall number, rdi=fd, rsi=buffer, rdx=length

Paradigm: Imperative

Typing: Untyped

Usage: Systems programming, embedded, OS kernels

Lines: 13

★ Every program you have ever run was ultimately executed as machine code. Assembly is as close as you can get to the hardware while still using human-readable text.

FLOW-MATIC 1955

Grace Hopper, Remington Rand ○ Extinct

Grace Hopper created FLOW-MATIC as the first programming language to use English-like keywords instead of mathematical notation. It has no Hello World, as a batch language, it had no terminal output, but its historical importance earns its place. Its syntax, with words like COMPARE, TRANSFER, and WRITE, directly inspired COBOL, proving that programming could read like English.

```
INPUT INVENTORY FILE-A
  COMPARE PRODUCT-NO (A) WITH PRODUCT-NO (B)
  IF GREATER GO TO OPERATION 10
  WRITE ITEM-A ON OUTPUT FILE-C
CLOSE-OUT FILES A B C
```

English-like syntax; no Hello World possible, FLOW-MATIC was batch-oriented

Paradigm: Imperative

Typing: Static

Usage: ●○○○○

Lines: 5

★ When Hopper first proposed using English words in code, her colleagues said it was impossible, computers could only understand math. She built FLOW-MATIC to prove them wrong.

IPL 1956

Allen Newell, Cliff Shaw, Herbert Simon ○ Extinct

Newell, Shaw, and Simon created IPL to build the Logic Theorist, one of the first AI programs. It has no Hello World, IPL was assembly-like, manipulating symbol lists rather than printing text; its list-processing concepts directly inspired LISP and shaped every AI language that followed. Its historical importance earns its place.

```
100 J150 SYMB 0 LINK 0
101 J151 SYMB 9-1 LINK 100
```

Assembly-like cell notation; no string output capability

Paradigm: Functional

Typing: Dynamic

Usage: ●○○○○

Lines: 2

★ The Logic Theorist, written in IPL, proved 38 of 52 theorems from Principia Mathematica, it was the first program to mimic human reasoning.

FORTRAN 1957

John Backus, IBM team ● Active

John Backus and his IBM team created FORTRAN (FORmula TRANslation) to make programming practical for scientific computing. It was transformative: fast, efficient, and powerful for real-world problems. FORTRAN showed that high-level languages matched hand-written machine code in speed, shattering the myth that efficiency required assembly language.

```
      PRINT *, 'Hello, World!'
      END
```

Column-based syntax (columns 7-72 for code)

Paradigm: Imperative

Typing: Static

Usage: ●●●○○

Lines: 2

★ FORTRAN remained in use for climate modeling and scientific computing well into the 2020s, some code from the 1960s ran production systems for over fifty years.

LISP 1958

John McCarthy, MIT ● Active

John McCarthy at MIT created LISP (LISt Processing) to transform artificial intelligence. His elegant insight was profound: code and data could both be represented as lists, enabling programs to manipulate themselves. This reflexivity enabled early AI systems and introduced functional programming to the world. McCarthy wanted to think mathematically about computation itself.

```
(format t "Hello, World!~%")
```

Lisp's fundamental syntax: (function args)

Paradigm: Functional

Typing: Dynamic

Usage: ●●○○○

Lines: 1

★ LISP's homoiconicity (code as data) was so ahead of its time that it influenced programming for 60+ years, it's fundamentally why modern metaprogramming exists.

COBOL 1959

CODASYL committee (led by Grace Hopper) ● Active

Grace Hopper led the CODASYL committee to create COBOL (COmmon Business-Oriented Language), specifically designed for business data processing. Hopper believed programming should be readable to non-technical managers, so COBOL used English-like syntax. Millions of businesses adopted it, making it the language of commerce for decades. It runs financial systems worth trillions of dollars.

```
IDENTIFICATION DIVISION.
PROGRAM-ID. HelloWorld.
PROCEDURE DIVISION.
    DISPLAY 'Hello, World!'.
    STOP RUN.
```

Highly structured; requires division headers

Paradigm: Imperative

Typing: Static

Usage: ●●○○○

Lines: 5

★ An estimated 220 billion lines of COBOL code remained running in banks, insurance, and government as of 2026, some written in 1965 and never touched since.

RPG 1959

IBM development team ◐ Endangered

An IBM team created RPG (Report Program Generator) for the IBM 1401, designed so business operators, not programmers, could generate reports from punch cards. Each column on the coding sheet had a fixed meaning, making it feel more like filling out a form than writing code. RPG evolved through decades of IBM hardware and remains embedded in thousands of IBM AS/400 and iSeries installations running payroll, inventory, and accounting.

```
     C     'Hello, World!'DSPLY
     C                    SETON                     LR
```

DSPLY displays a literal; SETON LR signals last record

Paradigm: Imperative

Typing: Static

Usage: ●○○○○

Lines: 2

★ RPG is so tied to IBM hardware that its community calls themselves 'RPGers', and IBM's midrange servers still ship with RPG as a first-class language, six decades after the punch-card era ended.

A SPOTLIGHT ON FLOW-MATIC AND COBOL

The Woman Who Made Computers Speak English

Grace Hopper, the Impossible Compiler, and COBOL's Sixty-Five-Year Reign

---

In 1952, Grace Hopper had a working compiler, the first program ever written that could translate human-readable instructions into machine code, and nobody would touch it.

"They told me computers could only do arithmetic," she recalled. "They patiently explained to me that computers could not do programs."

She spent the next three years proving them wrong, and the next three decades making sure they stayed wrong. The result was a language called COBOL that was supposed to be temporary, and a sixty-five-year reign that shows no sign of ending.

Grace Brewster Murray was born in New York City on December 9, 1906, into a family that prized curiosity. As a child, she dismantled so many household alarm clocks to understand their mechanisms that her parents limited her to one at a time. She graduated Phi Beta Kappa from Vassar, earned a PhD in mathematics from Yale in 1934, and was teaching when Pearl Harbor changed everything. She tried to enlist in the Navy but was rejected, at thirty-four, too old, and at fifteen pounds below the minimum, too small. She got an exemption, entered the Naval Reserve, and graduated first in her class from Midshipmen's School.

Her first assignment: the Mark I computer at Harvard. Fifty feet long, eight feet high, with 530 miles of internal wiring, the Mark I was the most complex calculating machine in the world. Hopper became one of its first three programmers and wrote its 561-page operating manual. Her team ran calculations for the Manhattan Project, numbers that went directly to John von Neumann.

After the war, Hopper joined the Eckert-Mauchly Computer Corporation and had the insight that would define her career. If a computer could manipulate mathematical symbols, she reasoned, it could manipulate words. In 1952, her team built the A-0 system, the first compiler. The establishment's reaction was blunt dismissal. Hopper spent years in what she called a "selling job," trying to convince anyone that the concept was even possible.

She persisted, developing FLOW-MATIC between 1955 and 1959, the first programming language to use English keywords like MULTIPLY and WRITE instead of mathematical notation. Her argument was characteristically direct: "Very few people are really symbol manipulators. If they are, they become professional mathematicians, not data processors."

The creation of COBOL began at a Pentagon meeting on May 28, 1959, when forty-one people from government, academia, and industry gathered to design a common business language. The politics were fierce. Hopper's FLOW-MATIC competed with IBM's COMTRAN. A committee member named Howard Bromberg, apparently fed up with the process, bought a fifteen-dollar tombstone, had COBOL engraved on it, and mailed it to the chairman. But FLOW-MATIC had a decisive advantage: it was the only business language actually implemented and running on real hardware. On December 6–7, 1960, the same COBOL program ran successfully on both an RCA and a UNIVAC computer, proving that a single program could work across different manufacturers' machines.

COBOL was designed as a stopgap. Everyone assumed it would be replaced within a few years. Instead, it became the backbone of civilization's financial infrastructure. By the 2020s, an estimated 95 percent of ATM transactions and 80 percent of in-person credit card swipes flow through COBOL code. Roughly 220 billion lines remain in active use worldwide.

The language's most darkly comic moment came in April 2020. As the pandemic triggered a flood of 362,000 unemployment claims in two weeks, New Jersey's forty-year-old COBOL mainframes buckled under the load. Governor Phil Murphy went on television to plead for volunteers who knew the language, mispronouncing it as "Cobalt." The programmers who answered the call were mostly over sixty, making them especially vulnerable to the very virus whose economic devastation they'd been summoned to manage.

Hopper herself became one of computing's great characters. She kept a clock on her wall that ran counterclockwise. She handed out pieces of wire cut to exactly 11.8 inches (the distance light travels in one nanosecond) to admirals, generals, and congresspeople, making the abstract tangible.

She retired from the Navy at seventy-nine as a rear admiral, the oldest active-duty officer in the United States Armed Forces, in a ceremony held aboard the USS Constitution, the oldest commissioned ship in the fleet. A guided-missile destroyer, a Cray supercomputer, a Yale residential college, and an Nvidia GPU architecture all bear her name.

Her most famous warning doubles as her epitaph: "The most dangerous phrase in the language is, 'We've always done it this way.'"

A SPOTLIGHT ON LOGO

The Exile's Classroom

Seymour Papert, Apartheid, and the Language That Taught Children to Think

Seymour Papert sat with a car transmission at his feet, a young boy in Pretoria mesmerized by the way gears turned against each other. He was perhaps ten years old when these machines became his first teachers, objects far more generous than the schools of apartheid South Africa, which had decided what knowledge was permissible based on the color of a child's skin. In his own home, working with those transmissions, Papert experienced something revolutionary: he was constructing understanding through play, learning mathematics through intimate contact with mechanical objects rather than through instruction. Many years later, when he tried to explain why he believed children learned best by building things, he would return to that moment. "Working with differentials did more for my mathematical development than anything I was taught in elementary school," he wrote. But it was his brother's geometry set and a compass and protractor, tools for measuring the world, that finally crystallized something profound: the objects were not separate from thinking. They were thinking, made tangible.

This is the man who would create Logo, the programming language that made the computer itself into a tool for thought rather than a machine for absorbing instruction. But before there was Logo, before there was MIT and the artificial intelligence laboratory and the famous turtle that children would command across screens and classroom floors, there was something more dangerous and more formative: a young man in an apartheid nation who had decided that the system's racial hierarchy, presented as inevitable, was breakable.

Seymour Aubrey Papert was born in Pretoria on February 29, 1928, into a Jewish family of Lithuanian descent at a moment when South Africa's racial architecture was still being cemented into law. As a child and teenager, he watched the apartheid system harden, separate schools, separate streets, separate futures assigned by skin color. He also watched something else: his own conscience refuse accommodation. By his university years at the University of the Witwatersrand, Papert had become a leading figure in anti-apartheid circles, organizing lessons for Black South Africans denied access to education, publishing in underground journals, speaking out against racial policy, placing himself deliberately in the path of a government that did not tolerate such resistance.

When he finally left South Africa in the 1950s, part self-exile, part necessity, it was with the understanding that he was fleeing a country that had marked him as troublesome. He went to Cambridge University for a second doctorate in mathematics, carrying with him the formative knowledge that hierarchies presented as natural are in fact human constructions, and human constructions can be dismantled.

In Geneva, from 1958 to 1963, Papert worked alongside Jean Piaget, the giant of developmental psychology who had transformed our understanding of how children learn. Piaget believed that children construct their own understanding rather than waiting, like vessels, to be filled with knowledge; that they learned by doing, by manipulating objects, by bringing their own theories to bear on the world and then revising those theories when reality contradicted them. To Papert, this was more than educational theory. It was liberation, the opposite of apartheid's passive transmission model, where authorities decided what should be learned and poured it into compliant students. Piaget's vision matched something Papert had already learned from gears and transmissions: that meaningful learning happens when a mind encounters something real and fights to understand it.

Papert extended Piaget's insight into what he would call constructionism: the idea that children learn most powerfully by constructing things, physical or computational objects, that embody the ideas they are constructing internally. A child who builds a sandcastle is learning about structure, weight, and failure. A child who writes a computer program is learning about logic, debugging, and the pleasure of making something work. The key insight was profound: the child should program the computer, not the other way around. The computer should not be a tool for programming the child into compliance.

In 1963, Papert came to MIT, joining the newly formed Artificial Intelligence Laboratory alongside Marvin Minsky. For four years he worked in a different kind of laboratory, where the question was how to understand human thought rather than control it, and perhaps how to make machines that could help people think better. The breakthrough came in 1967, in collaboration with Wally Feurzeig and Cynthia Solomon at Bolt, Beranek and Newman, a research firm in Cambridge, Massachusetts. They created Logo, a programming language that looked nothing like the complicated symbolic logic of LISP, which was what computer scientists used. Instead, Logo was simple enough for a child to grasp. FORWARD 50. RIGHT 90. BACKWARD 30. With these commands, a child could control a turtle, first an actual small robot that rolled across the floor, later a graphic cursor that drew on a screen, and watch as simple instructions produced visible results.

The turtle made Logo famous. The philosophy embedded in the language made it revolutionary. Logo emerged from Papert's understanding that children needed to be able to think about their own thinking, to debug their own programs, to encounter failure as a natural part of learning rather than a sign of inability. Every child who used Logo experienced the delight and frustration of giving a command, seeing the unexpected result, and then revising the code to get closer to intention. Nothing was being transmitted to a passive recipient; the child was constructing understanding, one revision at a time. In a sense, Logo carried forward what Papert had been trying to teach in those clandestine evening lessons decades earlier in Pretoria, the radical proposition that a person could learn, that learning was possible, that thinking was something everyone could do.

In 1980, Papert published Mindstorms: Children, Computers, and Powerful Ideas, a book that articulated the philosophy behind Logo with remarkable clarity and warmth. He argued that computers had been conscripted into the service of traditional schooling, used to present content more efficiently, to track students more thoroughly, to individualize the very standardization that made schooling oppressive. An alternative was possible. A child could use a computer the way a musician uses an instrument, as a medium for expressing and exploring ideas. The book found an audience hungry for what Papert was describing. Throughout the 1980s, Logo appeared in thousands of schools, and children sat down at terminals and learned what it felt like to be in command of technology rather than commanded by it.

Papert's vision was not naive about technology's power to harm. He had lived through apartheid; he knew that technology could be inscribed with power and used to enforce hierarchy. But he believed that in the right hands, especially in the hands of children learning to think for themselves, computers could be tools for liberation. Programming, in Papert's view, amounted to a form of literacy, a way of engaging with the world. When you debug code, you learn to be precise about language. When you design a program, you learn to think about structure. When you create something that works, you experience the power of your own mind in a way that no amount of passive instruction could provide.

For more than half a century, Papert carried this vision forward. His work influenced everyone from educators building classrooms around Logo to computer scientists like Mitchel Resnick, who would later create Scratch, a descendant of Logo that would introduce programming to millions of children worldwide.

In December 2006, while traveling to a conference in Hanoi, Papert was struck by a motorcycle while crossing a street. The accident caused a severe brain injury. He was seventy-eight years old. He worked on rebuilding his capacities with the same constructionist spirit that had animated his entire career, engaging actively in his own rehabilitation rather than passively receiving care.

Seymour Papert died on July 31, 2016, at the age of eighty-eight. In his wake, children around the world continued to learn by making things, by directing robots, by writing code. But perhaps the deepest legacy was no specific technology at all; it was the conviction that learning is not something done to people but something people do. That authority's greatest mistake is to assume it knows what a person is capable of understanding. In the exile of his early activism in apartheid South Africa, Papert had learned a lesson that shaped everything that followed: that the most powerful education is one that teaches people to think for themselves, to trust their own observations, to build their own understanding. Logo was simply the language he found to teach everyone else what gears had taught him.

What happened next

Year	Event
1928	Seymour Papert born in Pretoria, South Africa
1940s-50s	Becomes active in anti-apartheid movement; organizes lessons for Black South Africans denied education
1952	Earns PhD in mathematics from University of the Witwatersrand; later earns second PhD from Cambridge in 1959
1958	Begins work with Jean Piaget in Geneva, studies constructivist theory
1963	Joins MIT AI Lab, co-founded with Marvin Minsky
1967	Creates Logo with Wally Feurzeig and Cynthia Solomon at BBN
1980	Publishes Mindstorms: Children, Computers, and Powerful Ideas
1980s	Logo adopted in thousands of schools worldwide
2006	Struck by motorcycle in Hanoi, suffers severe brain injury
2016	Dies on July 31 at age eighty-eight

Era 2

The Classical Period

1960–1979

The 1960s and 1970s were programming's classical age, two decades in which the foundations of nearly every modern language were laid. International committees created ALGOL 60 to standardize algorithmic notation, and its block structure and formal grammar became the blueprint for languages that followed. SIMULA introduced classes and objects, planting the seed that would grow into the dominant paradigm of the next century. BASIC democratized programming by making it accessible to anyone with a computer terminal.

The 1970s brought a remarkable burst of creativity. Dennis Ritchie's C demonstrated that a small, simple language could be powerful enough for an entire operating system. Smalltalk showed that objects, message passing, and dynamic typing could be beautiful. Prolog proved that programs could be written as logical declarations rather than step-by-step instructions. SQL made databases queryable in something close to English. And practical tools emerged too, AWK for text processing, Scheme for elegant minimalism, the Bourne Shell for orchestrating Unix. By 1979, the foundations of modern programming were firmly in place: structured control flow, data types, functions as abstractions, object-oriented thinking, logic programming, and the beginnings of formal semantics.

21 languages in this era

Timeline

- 1960 ALGOL 60 published; becomes standard for algorithm publication
- 1962 Ole-Johan Dahl and Kristen Nygaard create SIMULA, introducing classes and OOP; Ralph Griswold creates SNOBOL
- 1964 John Kemeny and Thomas Kurtz create BASIC for Dartmouth Time-Sharing System
- 1966 Kenneth Iverson creates APL with mathematical symbols
- 1969 Ken Thompson creates B at Bell Labs
- 1970 Charles Moore creates Forth; Niklaus Wirth creates Pascal as the ideal teaching language
- 1972 Dennis Ritchie creates C; Alan Kay and Xerox PARC develop Smalltalk; Alain Colmerauer creates Prolog
- 1974 Barbara Liskov creates CLU at MIT; SQL developed at IBM
- 1975 Sussman and Steele create Scheme at MIT
- 1977 AWK created by Aho, Weinberger, Kernighan
- 1979 Stephen Bourne creates sh

World Context

Moon landing (1969); Vietnam War; Watergate; personal computing revolution beginning; Cold War tensions

Dominant Paradigms

Structured programming becomes standard; OOP emerging in Smalltalk; functional programming in Lisp/Scheme; logic programming with Prolog; Unix and C reshape systems programming

ALGOL 60

1960

International committee (Dijkstra, Naur, others)

○ Extinct

ALGOL 60 was created to improve on ALGOL 58, refined by legendary computer scientists including Edsger Dijkstra. By the 1960s, it was the standard for publishing algorithms in academic journals because of its clarity and expressiveness. ALGOL 60 introduced recursive functions and structured programming, foundational concepts that laid the groundwork for every modern language.

```
begin
  print('Hello, World!')
end
```

Block structure that shaped every language after it, begin/end became the universal pattern

Paradigm: Imperative

Typing: Static

Usage: ●○○○○

Lines: 3

★ ALGOL 60's influential report was so well-written that it set the standard for language documentation, computer scientists still study it as the model of technical clarity.

A SPOTLIGHT ON C

The Quiet Craftsman

Dennis Ritchie, Bell Labs, and the Language That Built Everything You Use

Dennis MacAlistair Ritchie was born in 1941 in Bronxville, New York, and inherited a computational world. His father, Alistair E. Ritchie, was a Bell Labs scientist who had co-authored "The Design of Switching Circuits," one of the foundational texts on switching theory. The son would follow the father to the same building, but not before wandering through physics.

At Harvard, Ritchie studied physics and applied mathematics, concentrating in physics as an undergraduate. He went on to graduate work in applied mathematics, completing a draft PhD thesis on computational complexity under the supervision of Patrick C. Fischer. But something shifted in him during those years, a kind of intellectual humility that would later define his work. "My undergraduate experience convinced me that I was not smart enough to be a physicist, and that computers were quite neat," he reflected. "My graduate school experience convinced me that I was not smart enough to be an expert in the theory of algorithms and also that I liked procedural languages better than functional ones." It was the sort of self-deprecation that often masks clarity of vision. By dismissing what he was not, he had already chosen what he would become.

In 1967, Ritchie joined Bell Labs' Computing Science Research Center as a young researcher. In the late 1960s, he and Ken Thompson worked on Multics, the ambitious time-sharing operating system that sought to democratize access to computing power. The work was noble. It was also, by Bell Labs' calculation, not quite paying off. When the laboratory pulled out of the Multics project in 1969, Thompson and Ritchie did not see failure. They saw an opportunity to start over.

Thompson found an old PDP-7 minicomputer, machines so small and underpowered by modern standards that they seem almost like found objects. On this machine, starting in 1970, he and Ritchie began to build something new: an operating system they called Unix, a self-aware pun on Multics' pretensions to universality. But Thompson had a problem. Assembly language, the raw machine tongue, was the only way to implement an operating system in those days. Assembly was precise, relentless, and utterly unportable. Write it for one machine, and you had rewritten nothing for any other. This was not sustainable. So Thompson created B, a language derived from BCPL, that sat above the machine but close enough to touch its wires.

This is where Ritchie entered the act. Between 1971 and 1973, he took Thompson's B and evolved it into C. The step was subtle but revolutionary. Ritchie added types, integers, characters, floating-point numbers, and user-defined structures. He added operators, refined the syntax, and created a system where the language could speak precisely to the machine's hardware while remaining human-readable enough that complex logic could be expressed without drowning in assembly mnemonics. The philosophy underneath was radical in its simplicity: trust the programmer. If someone wanted to cast a pointer or manipulate memory directly, C would let them. The language assumed competence. It did not hold your hand, and it did not protect you from yourself. This was not a bug. This was the feature. By 1973, Ritchie and Thompson had rewritten Unix in C. It was the first time a major operating system had been written in a high-level language. That sentence does not sound particularly earth-shaking in an era when all operating systems are written in high-level languages, but in 1973 it violated the received wisdom of computer science. The machines could not be trusted to understand humans. You had to speak to them in their native dialect, in assembly. Ritchie proved that was wrong. More than that, he proved that you could write an operating system in a high-level language, make it portable, and watch it spread to every computer in the world.

In 1978, Ritchie and Brian Kernighan published "The C Programming Language", K&R, as it came to be known. The book was part reference manual, part primer, with Kernighan's clear expository prose wrapped around Ritchie's authoritative definition of the language. It became the bible. For years, there was no official standard for C. K&R was C. The book sold millions of copies and made the language a household name in computer science departments, research labs, and the bedrooms of obsessive programmers. The book's influence cannot be overstated. It was documentation that doubled as an argument about how to think about computing: precisely, pragmatically, without fuss. Ritchie himself was precise and pragmatic, and without fuss. He rarely gave interviews. He avoided the spotlight. When Unix and C began to dominate computing in the 1980s and 1990s, conquering server rooms and embedded systems and the infrastructure of the internet itself, Ritchie did not become a celebrity. He remained at Bell Labs, later part of Lucent Technologies, working quietly. In 1983, he and Thompson received the A.M. Turing Award, the field's highest honor, for their development of Unix. The citation noted that "The success of the UNIX system stems from its tasteful selection of a few key ideas and their elegant implementation." That phrase, "tasteful selection of a few key ideas," could have been written about Ritchie himself. He had a gift for knowing what to include and what to leave out.

The impact of his work is almost impossible to quantify. Everything you use runs on Unix or an operating system descended from it. The servers that run the internet. The phones in your pocket. The ATMs that dispense your money. The medical devices keeping people alive. And nearly all of it, directly or indirectly, is written in C or in languages that descend from C's design principles. Python owes C its syntax. JavaScript owes C its expression. Even Rust and Go, languages built explicitly to be safer and more modern than C, are reactions to C, arguments with C about what programming should be.

But Ritchie did not live to see the full flowering of his legacy. In October 2011, he died at his home in Berkeley Heights, New Jersey, at the age of seventy. The news came almost silently. Steve Jobs had died a week earlier, and the technology world had erupted in mourning for the man who made computers beautiful. Ritchie's death went largely unnoticed. Rob Pike, who had worked alongside Ritchie for twenty years at Bell Labs, wrote on Google+ simply: "Dennis was my friend, colleague, and collaborator, and the world has lost a truly great mind. His influence on today's world was enormous." Pike had to remind people that Ritchie existed, that he mattered, that the world runs on his work. Most did not even know to mourn.

This is the way of quiet craftsmen. They do not seek the stage. They make things that last, and then they disappear. They leave behind code that keeps running, languages that keep evolving, operating systems that keep serving. Their names fade. Their work endures. Dennis Ritchie spent his life at Bell Labs, in Room 2C-517 in Murray Hill, New Jersey, building the foundations that the digital world rests upon. He did it without fanfare, without corporate ambition, without the hunger for recognition that drives so many brilliant people. He did it because he thought computers were quite neat. And because he was smart enough to understand what the world would need.

What happened next

Year	Event
1967	Dennis Ritchie joins Bell Labs Computing Science Research Center
1969	Bell Labs withdraws from Multics project; Thompson and Ritchie begin Unix development
1970	Development of Unix begins on PDP-7 minicomputer
1971–1973	C language designed and refined; Unix rewritten in C

1978	"The C Programming Language" (K&R) published
1983	Ritchie and Thompson awarded A.M. Turing Award
1993	Ritchie publishes "The Development of the C Language" at HOPL II
2011	Dennis Ritchie dies October 12 in Berkeley Heights, New Jersey

SIMULA 1962

Ole-Johan Dahl, Kristen Nygaard, Norway ● Active

Norwegian scientists Ole-Johan Dahl and Kristen Nygaard created SIMULA to simulate real-world systems, originally for naval warfare scenarios. They introduced the groundbreaking concept of 'classes' and object-oriented thinking to programming. SIMULA showed that OOP was a way to model reality itself, not merely a coding style, directly inspiring Smalltalk and modern OOP.

```
BEGIN
  OUTTEXT('Hello, World!');
  OUTIMAGE
END
```

OUTTEXT *outputs strings,* OUTIMAGE *outputs line*

Paradigm: Object-Oriented

Typing: Static

Usage: ●○○○○

Lines: 4

★ SIMULA's object system was so refined that when Smalltalk designer Alan Kay saw it, he credited Dahl and Nygaard as the true fathers of object-oriented programming.

SNOBOL 1962

Ralph Griswold, Bell Labs ○ Extinct

Ralph Griswold at Bell Labs created SNOBOL (StriNg Oriented symBOlic Language) to advance string processing and pattern matching. In an era when computers were seen as number-crunchers, Griswold believed they should manipulate text. SNOBOL was the language of choice for linguists, text processors, and anyone doing symbolic manipulation.

```
OUTPUT = 'Hello, World!'
END
```

Simple assignment to OUTPUT variable

Paradigm: Imperative

Typing: Dynamic

Usage: ●○○○○

Lines: 2

★ SNOBOL's pattern matching was so advanced that it influenced every regex engine that followed, every time you use grep or regex, you're benefiting from Griswold's 1962 innovation.

BASIC 1964

John Kemeny, Thomas Kurtz, Dartmouth ● Active

John Kemeny and Thomas Kurtz at Dartmouth created BASIC (Beginner's All-purpose Symbolic Instruction Code) to democratize computing. They believed every student should learn to program, so they designed a language anyone could understand in minutes. BASIC was the gateway drug to programming for millions of hobbyists and students, fueling the personal computer boom.

```
10 PRINT "Hello, World!"
20 END
```

Line numbers required; PRINT statement

Paradigm: Imperative

Typing: Dynamic

Usage: ●●○○○

Lines: 2

★ BASIC's simplicity made it the first language on home computers like the Commodore 64 and Apple II; it directly created the 1980s gaming industry.

PL/I

1964

IBM

◐ Endangered

IBM created PL/I (Programming Language One) as an ambitious 'universal' language combining features of COBOL, FORTRAN, and ALGOL. IBM wanted one language for all computing: scientific, business, systems. PL/I was feature-rich but notoriously complex, earning a reputation for feature creep and difficult debugging.

```
Hello: proc options(main);
  put list ('Hello, World!');
end Hello;
```

Procedure declaration with options(main); put list outputs

Paradigm: Imperative

Typing: Static

Usage: ●○○○○

Lines: 3

★ PL/I was so feature-rich that the language specification was over 1,000 pages, a case study in feature creep's diminishing returns.

APL 1966

Kenneth Iverson, IBM ◐ Endangered

Kenneth Iverson at IBM created APL (A Programming Language) based on his mathematical notation system. APL replaced English keywords with mathematical symbols, allowing mathematicians to express algorithms with extreme conciseness. Physicists and mathematicians loved it for its power, though its symbols mystified everyone else.

```
a←'Hello, World!'
a
```

The left arrow assigns; typing a variable name prints its value

Paradigm: Functional

Typing: Dynamic

Usage: ●○○○○

Lines: 2

★ APL code was so dense that a single line could accomplish what took a page of C; programmers called it 'write-only' because your own code would be incomprehensible weeks later.

A SPOTLIGHT ON SEQUEL AND SQL

The Last Thing They Did Together

Don Chamberlin, Raymond Boyce, and the Language That Runs Most of the World's Databases

---

The carpool from the IBM Research Laboratory in San Jose was routine. Don Chamberlin drove that day in June 1974, the same route he'd taken countless times with his best friend and colleague Raymond Boyce. They were coworkers and neighbors in the suburban sprawl south of the Bay, the kind of friends who built their lives in proximity, who debated query languages over morning coffee and Saturday dinners. They'd moved to California together, left the East Coast behind, and thrown their considerable minds at a single problem: how to make Edgar Codd's theoretical revolution into something real, something practical, something that ordinary people could actually use.

By lunchtime that day, Boyce had collapsed in the cafeteria. A brain aneurysm. No warning. No goodbye. He died on June 18, 1974, leaving behind his wife Sandy and their infant daughter Kristin. He was twenty-seven years old. He and Chamberlin had just published their first SEQUEL paper at the ACM SIGFIDET conference in Ann Arbor, Michigan, less than a month earlier. That paper, delivered in May, already distributed, already being read by the people who would build the future, became the last thing they did together.

There is no single visionary genius in this story. There are two people who understood something that almost everyone else, including IBM itself, did not yet see.

In 1970, Edgar Codd, a British RAF veteran who had served as a pilot during the Second World War, published the paper that started all of it. "A Relational Model of Data for Large Shared Data Banks" appeared in Communications of the ACM. Codd had moved to IBM's San Jose Research Laboratory in 1968 and spent the next two years crystallizing decades of thought into eleven pages that would alter the course of computing. He proposed that data could be organized into simple tables, that these tables could be related to one another through common values, and that this organization could be described with mathematical precision. The model was elegant, revolutionary, and, in the eyes of the corporation paying for it, not ready for the market.

IBM had a successful database product called IMS, and was not in a hurry to replace it. Instead of rushing Codd's ideas to market, IBM organized a research project, called System R, to investigate the feasibility of relational databases. The research project was allowed, even encouraged, to publish its work in the open technical literature. A professor at Berkeley named Michael Stonebraker would later wonder aloud why IBM, having invented the relational model, had taken so long to commercialize it.

In the summer of 1973, Don Chamberlin and Raymond Boyce arrived in San Jose to work on the System R project. Their mandate was to implement Codd's model. They needed a language for ordinary people, for database administrators and office workers and anyone who simply wanted to ask the database a question, not one for mathematicians writing predicate calculus. They called it SEQUEL: Structured English Query Language. It read almost like English. A person could learn it. A person without a PhD in mathematics could sit down and write a query.

The two men had developed their ideas by playing what they called "the query game." Chamberlin would dream up a question he wanted to ask of a database, and Boyce would write the SEQUEL code to answer it. Then they'd reverse roles. Question, answer. Answer, question. Back and forth, refining the language, simplifying it, testing the boundaries of what could be expressed, learning from each attempt what the language still needed. None of it was academic exercise. It was craftsmanship, two people treating a programming language as something that should serve human beings, not the other way around.

In early May 1974, in Ann Arbor, Michigan, they presented "SEQUEL: A Structured English Query Language" to the ACM SIGFIDET workshop. The paper was straightforward, practical, almost modest in tone. Here is what we built. Here is how it works. Here is why we think people should use it instead of wrestling with complex operations. Within a month, Boyce was dead. Chamberlin, who had relied on his friend for the intellectual partnership that had powered their work, had to absorb the loss and continue alone.

But there was a young entrepreneur paying attention. Larry Ellison read the System R papers and correctly predicted that IBM would eventually release a SEQUEL database on its mainframe computers. Seeing a market opportunity, Ellison built his own SEQUEL database on a less expensive minicomputer.

Ellison's product, which he named Oracle, beat IBM to the relational database market by about two years.

The language itself would not remain unchanged. Trademark disputes (a company called Hawker Siddeley owned the rights to SEQUEL) forced a renaming. They dropped the vowels. SEQUEL became SQL. The acronym stood for Structured Query Language now, not Structured English Query Language, and though Chamberlin would spend the rest of his life pronouncing each of those three letters when others said it as a word, that change was pragmatic and necessary. It had to survive. The language had to survive.

And it did. Chamberlin remained at IBM, elected eventually to the National Academy of Engineering. He watched as SQL became the most widely used database query programming language on Earth. By most counts, more queries are executed in SQL every day than in any other language, across more machines, more systems, more databases than any other framework for asking questions of stored information. Banks run on it. Hospitals run on it. Airlines run on it. Every purchase you make online runs through SQL. Every balance inquiry, every reservation, every transaction that matters to modern life has probably been processed by SQL code.

Boyce did not live to see this. But his name endured. While working on the theoretical foundations of relational databases, he had independently articulated what became known as Boyce-Codd Normal Form, a mathematical tool for designing databases properly, for eliminating redundancy, for ensuring data integrity. It bears his name still. In database schema design classes all over the world, students learn about BCNF, and they're learning the contribution of a man who died at twenty-seven and never saw the full impact of his work.

There is something almost unbearable about this. Two friends, neighbors, carpool partners, devoted to making something powerful also usable. One of them dies after shipping the thing they built together. The other continues in solitude, carrying the work forward, carrying the loss, watching as the language they created together becomes central to electronic commerce, the foundation of everything. That final paper, delivered together in May 1974, stands as the moment when they were both still here, still building, still believing that clarity and accessibility could be expressed through mathematics and code.

SQL was not the only language born from that era. But it may be the only one that became truly indispensable. It did not earn that place by being the most elegant, or by winning some theoretical competition. It earned it because Chamberlin and Boyce built it for human beings instead of for computers. They asked a fundamental question: how would a person naturally want to ask a database for information? And then they built a language around the answer.

The language outlasted them both, though Chamberlin lived to see it triumph and Boyce did not. Today, fifty years later, when a database administrator needs to retrieve data, when a data scientist needs to explore a dataset, when a company needs to query the millions of customer records on which its business depends, they use SQL. They use the last thing Don Chamberlin and Raymond Boyce did together. They use it dozens of times a day. They use it without thinking about the two friends in California who drove to work together and changed everything, or the one who came home alone.

What happened next

Year	Event
1970	Edgar Codd publishes "A Relational Model of Data for Large Shared Data Banks," IBM largely ignores it; Larry Ellison reads the paper and begins planning Oracle
1973	IBM organizes System R project at San Jose Research Lab. Chamberlin and Boyce join the System R project at IBM Research Lab, San Jose; they develop SEQUEL through iterative design and the "query game"
May 1974	Chamberlin and Boyce present "SEQUEL: A Structured English Query Language" at ACM SIGFIDET conference in Ann Arbor, Michigan
June 18, 1974	Raymond Boyce dies from a brain aneurysm at age 27; Chamberlin drives him to work that morning
1975	SEQUEL renamed to SQL due to trademark conflict with Hawker Siddeley aircraft company; continued development proceeds under new name
1978	IBM's System R prototype proves SQL commercially viable; leads to SQL/DS and DB2
1979	Oracle Version 2 becomes the first commercially available relational database product using SQL, beating IBM to market
1981	IBM releases its first SQL product, SQL/DS, followed by DB2 in 1983. Many other companies release SQL products throughout the 1980s
1986	SQL becomes ANSI standard; adopted as international standard by ISO in 1987

2024	Don Chamberlin receives recognition for 50 years of SQL; the language remains the most widely used database query language on Earth

MUMPS 1966

Neil Pappalardo, Robert Greenes, Curt Marble ● Active

Neil Pappalardo and colleagues at Massachusetts General Hospital created MUMPS (Massachusetts General Hospital Utility Multi- Programming System) to handle medical records on limited hardware. Its built-in database and terse syntax made it ideal for healthcare, where rapid lookups across millions of patient records mattered more than clean code. MUMPS remained embedded in hospital systems worldwide, managing life-or-death data.

```
WRITE "Hello, World!"
```

Single WRITE command; MUMPS uses a built-in hierarchical database

Paradigm: Imperative

Typing: Typeless

Usage: ●○○○○

Lines: 1

★ MUMPS powers VA VistA, the electronic health record system serving over 9 million US veterans, one of the largest healthcare IT systems ever built, running on a language most programmers have never heard of.

BCPL 1967

Martin Richards, Cambridge ○ Extinct

Martin Richards at Cambridge created BCPL (Basic Combined Programming Language) to simplify CPL's overwhelming complexity. BCPL was a practical, minimal language that proved a small set of features could be very potent. BCPL directly inspired C, which became the most influential systems programming language ever.

```
GET "LIBHDR"
LET START() BE $(
  WRITES("Hello, World!*N")
$)
```

GET includes standard header; WRITES outputs a string

Paradigm: Imperative

Typing: Weakly Static

Usage: ●○○○○

Lines: 4

★ BCPL's minimalism directly led to C's design; every feature in BCPL was questioned with 'Is this truly necessary?', the same philosophy that made C legendary.

LOGO

1967

Seymour Papert, MIT

● Active

Seymour Papert at MIT created Logo to teach children to think computationally. Based on LISP, Logo introduced the 'turtle', a simple robot that moved on a screen following commands. Papert believed programming could teach children how to think, transforming computer science education. Millions of children first learned programming through the turtle.

```
to hello
print [Hello, World!]
end
hello
```

Procedure definition with 'to' keyword

Paradigm: Functional

Typing: Dynamic

Usage: ●●○○○

Lines: 4

★ The Logo turtle sparked the entire field of educational technology; Papert's insight that children could learn to think through programming transformed education.

B 1969

Ken Thompson, Dennis Ritchie, Bell Labs ● Active

Ken Thompson created B at Bell Labs as a stripped-down version of BCPL for the PDP-7 minicomputer, which had only 8KB of memory. B was typeless (everything was a machine word) which made it fast but dangerous. Dennis Ritchie soon found B too limited and evolved it into C, making B the missing link in one of computing's most important lineages: BCPL to B to C.

```
main() {
  putstr("Hello, World!*n");
}
```

*Typeless; *n is newline (not backslash-n); predecessor to C*

Paradigm: Imperative

Typing: Typeless

Usage: ●○○○○

Lines: 3

★ B's name is simply the first letter of BCPL. When Ritchie created its successor, he called it C, the next letter. Programming history's most consequential alphabet lesson.

FORTH 1970

Charles Moore ◐ Endangered

Charles Moore created Forth as an ultra-minimalist language for controlling telescopes. Moore believed programming should be transparent and efficient, without abstraction layers hiding the machine. Forth used a stack-based approach with almost no syntax, enabling incredibly fast execution on minimal hardware. Astronomers and embedded systems programmers loved it.

```
: hello ." Hello, World!" CR ;
hello
```

Stack-based; ." " outputs string, CR is newline

Paradigm: Imperative

Typing: Dynamic

Usage: ●○○○○

Lines: 2

★ Forth was so efficient that it emerged as the standard for controlling telescope arrays and spacecraft; it found lasting use in aerospace because it ran on minimal computing power.

PASCAL 1970

Niklaus Wirth, Swiss Federal Institute ◐ Endangered

Niklaus Wirth at the Swiss Federal Institute created Pascal as the ideal teaching language, combining ALGOL's clarity with practical features. Wirth believed well-crafted language design could teach good programming practices. Pascal dominated computer science education for 20 years, producing generations of well-trained programmers.

```
program HelloWorld;
begin
  WriteLn('Hello, World!')
end.
```

Structured with begin-end blocks

Paradigm: Imperative

Typing: Static

Usage: ●●○○○

Lines: 4

★ Pascal's insistence on structured programming and clear syntax shaped an entire generation of programmers, many still consider it the most concise teaching language.

C 1972

Dennis Ritchie, Bell Labs ● Active

Dennis Ritchie at Bell Labs created C as a minimal, powerful systems programming language for writing Unix. Ritchie wanted a language that let programmers write fast code close to the machine without wrestling with assembly language. C achieved what seemed impossible: high-level expressiveness without sacrificing efficiency, making it the most influential language in computing history.

```
#include <stdio.h>
int main() {
  printf("Hello, World!\n");
  return 0;
}
```

Requires stdio.h; printf with format string

Paradigm: Imperative

Typing: Static

Usage: ●●●●●

Lines: 5

★ C is so foundational that Unix, Linux, Windows (kernel), macOS, Android, and virtually every operating system are written in C; it's the lingua franca of systems programming.

The Curly Brace

How a Character Set Limitation on a 1960s Machine Became the Most Ubiquitous Symbol in Programming

In the early 1960s, the University of Cambridge and the University of London built CPL, the Combined Programming Language, as an ambitious successor to ALGOL 60. It was supposed to be everything: powerful enough for systems programming, elegant enough for research, expressive enough for any problem. Christopher Strachey, David Barron, and others poured years into it. CPL was all of those things. It was also enormous, difficult to implement, and never completed. By the mid-1960s, it existed mostly as a specification, not a working tool. A failure that seeded a dynasty.

Martin Richards had spent three years as a research student on the CPL project, attending the design meetings, watching the complexity accumulate. In December 1966, he arrived at Project MAC, a department of MIT, with a question that the CPL experience had made urgent: what if you kept only the essential parts? What if a systems programming language could be so small that one person could implement it in weeks?

The only machine available was an IBM 7094. Richards chose D.T. Ross's AED0, one of the few languages available that supported the recursive functions he needed for his recursive descent parser. He began implementing BCPL in AED0. By about February 1967, he had a prototype working.

Then came the constraint that would echo through sixty years of programming. The IBM 7094 had a 36-bit word with 6-bit characters packed six per word. The character set was restrictive. It did not include { or }. So Richards chose $(and $) as his block delimiters instead. As soon as the AED0 version worked, he rewrote the compiler in BCPL itself, producing a compiler that could compile itself. The development was done on the CTSS time-sharing system, through an IBM golf ball typewriter connected to the mainframe. The typewriter could display a much larger character set than the 7094's native encoding. Before long, Richards switched to 9-bit characters packed four per word, which gave him access to that larger set. He added { and } as synonyms for $(and $).

That decision, adding curly braces as block delimiters because a better terminal made them available, would turn out to be one of the most consequential punctuation choices in the history of technology.

BCPL was typeless. Not weakly typed. Typeless. Every variable was a machine word, a bit pattern. The language did not know or care what that bit pattern represented. The operation you applied to it, integer addition, pointer dereference, character comparison, determined the interpretation. A CPU register works the same way: the accumulator does not have a type. It is just bits. BCPL treated all data with the same indifference.

This radical simplicity was the point. Where CPL had tried to be everything, BCPL tried to be just enough. It had functions, recursion, and a clean block structure. It compiled efficiently. It could be implemented quickly on new machines. And its minimalism asked a question that would define the next three languages in the lineage: what is the smallest set of features a systems programming language actually needs?

In 1969, Ken Thompson at Bell Labs needed a language for the PDP-7 minicomputer. The PDP-7 had 8KB of memory. BCPL, small as it was, was still too large. Thompson stripped it down further, removing features until only the essentials remained. He called the result B, the first letter of BCPL. B was even more minimal than its parent: still typeless, still treating everything as a machine word, but now small enough to fit on a machine the size of a refrigerator. Thompson used B to write utilities for a new operating system he and Dennis Ritchie were building. They called it Unix.

But B had a problem. The PDP-11, Bell Labs' next machine, had different-sized data types: bytes, words, and longer values. A typeless language could not express these distinctions. Between 1971 and 1973, Ritchie took Thompson's B and added what BCPL and B had deliberately left out: types. Integers. Characters. Floating-point numbers. Structures. The result was C. The curly braces that Richards had added to BCPL as synonyms, the ones made possible by a better terminal on an IBM 7094, came through B into C unchanged.

The alphabet tells the story: BCPL, B, C. Each language a reduction and refinement of the one before. Richards provided the block structure and the braces. Thompson provided the ruthless minimalism. Ritchie provided the types that made it practical for real hardware. Three people, three languages, five years, and the syntax of modern programming was set.

C's syntax became the template. C++ adopted it. Java adopted it. JavaScript adopted it. C#, Go, Rust, Swift, Kotlin, TypeScript: every curly-brace language traces its block-delimiter syntax back through the same lineage. When a programmer anywhere in the world opens a function body today, the character they type was put there by Martin Richards in 1967 because an IBM golf ball typewriter could finally display it. Not every language followed. Python chose indentation. Haskell chose layout rules. Ruby chose keywords. But the curly brace remains the default, the syntax that most working programmers in the world read and write every day. It started as a synonym.

The most consequential syntax in programming began as a workaround for a character set that could not display the symbol its creator wanted. The curly brace carried all three of their decisions into the future.

What happened next

Year	Event
1963	Christopher Strachey, David Barron, and others publish the first paper describing CPL; the language is being implemented at Cambridge and London but proves too ambitious to complete
1966	Martin Richards arrives at Project MAC, MIT, in December to build a practical successor to CPL
1967	Richards implements BCPL on an IBM 7094 using AED0; the 6-bit character set forces $(and $) as block delimiters; after switching to 9-bit characters, Richards adds { and } as synonyms
1967	Richards rewrites the BCPL compiler in BCPL itself, creating a self-hosting compiler on the CTSS time-sharing system
1969	Ken Thompson at Bell Labs strips BCPL down to create B for the PDP-7; Thompson and Ritchie begin building Unix
1971–73	Dennis Ritchie evolves B into C, adding types; the curly braces from BCPL carry through unchanged
1973	Ritchie and Thompson rewrite Unix in C, the first major operating system written in a high-level language

1978	Kernighan and Ritchie publish The C Programming Language; the curly-brace syntax becomes the industry standard
1983	Bjarne Stroustrup releases C++, extending C with classes; curly braces define every C++ block
1995	Java and JavaScript both adopt curly-brace syntax; the pattern spreads to the web
2000s–2020s	C#, Go, Rust, Swift, Kotlin, TypeScript, and others continue the curly-brace tradition

PROLOG 1972

Alain Colmerauer, Marseille University ● Active

Alain Colmerauer at Marseille University created Prolog (PROgramming LOGic) to advance logic programming. Instead of telling the computer what to do, Prolog lets you declare facts and rules, asking the computer to prove statements. This 'declarative' approach inspired artificial intelligence research and became the language of choice for knowledge representation.

```
:- initialization(main).
main :- write('Hello, World!'), nl, halt.
```

Declarative with initialization directive

Paradigm: Logic/Declarative

Typing: Dynamic

Usage: ●○○○○

Lines: 2

★ Prolog was the foundation of the Japanese 'Fifth Generation Computer' project, governments bet billions on Prolog as the future of AI.

SMALLTALK 1972

Alan Kay, Dan Ingalls, Xerox PARC ◑ Endangered

Alan Kay and Dan Ingalls at Xerox PARC created Smalltalk as a fully object-oriented system where everything is an object, even numbers and classes. Inspired by Simula 67, Kay pushed OOP to its ultimate conclusion. Smalltalk introduced the visual IDE, graphical programming environment, and made OOP tangible and formidable.

```
Transcript show: 'Hello, World!'
```

Message passing: 'show:' message to Transcript object

Paradigm: Object-Oriented

Typing: Dynamic

Usage: ●○○○○

Lines: 1

★ Smalltalk introduced the modern IDE with windows, menus, and graphical code editing ; the Xerox Alto was so advanced it predicted personal computing by years.

ML 1973

Robin Milner, Edinburgh ● Active

Robin Milner at University of Edinburgh created ML (Meta Language) to automatically prove theorems using the LCF proof assistant. ML introduced strong static typing with type inference, the compiler could figure out types automatically. This innovation became foundational for functional programming and influenced every statically typed language that followed.

```
val () = print "Hello, World!"
```

Type inference with print function

Paradigm: Functional

Typing: Static with Inference

Usage: ●●○○○

Lines: 1

★ ML's type inference system inspired Haskell, OCaml, and TypeScript; every modern language with smart type systems owes ML a debt.

CLU

1974

Barbara Liskov, MIT

○ Extinct

Barbara Liskov created CLU at MIT to explore data abstraction, the idea that a program should interact with data through defined interfaces, not by poking at its internals. CLU introduced iterators, abstract data types, and exception handling, concepts that became standard in Java, Python, C++, and nearly every modern language. Liskov later won the Turing Award partly for this work.

```
start_up = proc ()
  stream$putl(stream$primary_output(),
              "Hello, World!")
  end start_up
```

CLU uses $ for type operations; stream$putl prints a line

Paradigm: Object-Oriented

Typing: Static

Usage: ●○○○○

Lines: 4

★ CLU's influence is invisible but everywhere: every time you write a for-each loop, use an iterator, or catch an exception, you're using ideas Barbara Liskov pioneered in 1974.

SQL 1974

Donald Chamberlin, Raymond Boyce, IBM ● Active

Donald Chamberlin and Raymond Boyce at IBM created SQL (Structured Query Language) to simplify database access based on Ted Codd's relational model. SQL is a powerful declarative language that transformed data access: instead of writing procedural code to fetch data, users simply declared what data they wanted. SQL became the universal language of data, used by billions of applications.

```
SELECT 'Hello, World!';
```

SELECT statement; some databases require FROM

Paradigm: Declarative

Typing: Static

Usage: ●●●●●

Lines: 1

★ Raymond Boyce died of a brain aneurysm at age 26, just months after co-inventing SQL. He never saw his creation become the most widely used database language on Earth.

SCHEME 1975

Gerald Jay Sussman, Guy Steele, MIT ● Active

Gerald Jay Sussman and Guy Steele at MIT created Scheme as a minimal, elegant dialect of Lisp. They wanted to understand lambda calculus and functional programming at a deep level, so they designed a language with minimal semantics and maximum expressiveness. Scheme emerged as the favorite language for teaching computer science at top universities.

```
(display "Hello, World!") (newline)
```

Functional with display and newline

Paradigm: Functional

Typing: Dynamic

Usage: ●●○○○

Lines: 1

★ Scheme was used to teach SICP (Structure and Interpretation of Computer Programs), MIT's legendary computer science course that shaped generations of brilliant programmers.

AWK 1977

Alfred Aho, Peter Weinberger, Brian Kernighan, Bell Labs ● Active

Alfred Aho, Peter Weinberger, and Brian Kernighan at Bell Labs created AWK as a text processing language that became a Unix powerhouse. AWK made it trivial to write one-liners that processed gigabytes of log files, extracting patterns and computing statistics. Every Unix system administrator became an AWK expert out of necessity.

```
BEGIN { print "Hello, World!" }
```

BEGIN block executes before any input

Paradigm: Imperative

Typing: Dynamic

Usage: ●●○○○

Lines: 1

★ AWK is so efficient that system administrators still use 30-year-old AWK one-liners on billion-dollar infrastructure, it's the Swiss Army knife of text processing.

BOURNE SHELL / SH 1979

Stephen Bourne, Bell Labs ● Active

Stephen Bourne at Bell Labs created the Bourne shell as the standard command interpreter for Unix. It introduced key features like pipes, background jobs, and scripting capability. The Bourne shell became the de facto standard across Unix systems, and its syntax directly influenced every shell that followed.

```
#!/bin/sh
echo "Hello, World!"
```

Shebang line; echo outputs to stdout

Paradigm: Imperative

Typing: Dynamic

Usage: ●●●○○

Lines: 2

★ The Bourne shell's design philosophy, simple tools connected by pipes, became Unix's fundamental principle, transforming how software is built.

A SPOTLIGHT ON BASIC

4 AM in College Hall

BASIC and the Bet That Everyone Should Code

In 1960, computing had a gatekeeping problem. If you wanted to run a program at most universities, you punched your code onto cards, handed them to an operator behind a window, and waited, hours, sometimes days, to learn whether you'd misplaced a comma. The machines cost millions. The priesthood of trained operators guarded them jealously. And if you were an English major or a history student, no one thought you had any business touching a computer at all.

Two mathematicians at Dartmouth College thought this was absurd.

John George Kemeny had lived several lives before he turned thirty. Born in Budapest in 1926, he fled Hungary with his family in 1940 as anti-Jewish laws tightened. His grandfather, who refused to leave, was murdered in the Holocaust. At Princeton, Kemeny was drafted and sent to Los Alamos, where he worked as a human "computer" under Richard Feynman, performing calculations on bookkeeping machines where simple math could take a week. After the war, he returned to Princeton and became Albert Einstein's mathematical assistant at the Institute for Advanced Study. He earned his PhD under Alonzo Church, the father of lambda calculus, and was a full professor at Dartmouth by 27.

His partner, Thomas Kurtz, arrived at Dartmouth in 1956, a statistician from Illinois with a gentle manner and a radical conviction. "Neither John nor I were computer scientists," Kurtz later said. "We were mathematicians." But they shared an idea considered heretical at the time: every student, regardless of major, should have access to computing.

The existing languages were part of the problem. Kemeny and Kurtz looked at FORTRAN and saw a syntax designed to intimidate. They wanted something a first-year student could read as English: FOR I = 1 TO 10 STEP 2. No ambiguity. No priesthood required.

They called it BASIC: Beginner's All-purpose Symbolic Instruction Code. Then they recruited about a dozen undergraduates who worked fifty hours a week (on top of their coursework) to build both the Dartmouth Time-Sharing System and the BASIC compiler in roughly a year.

The moment arrived at 4:00 AM on May 1, 1964, in the basement of College Hall. Kemeny sat at one teletype terminal. A student programmer named John McGeachie sat at another. They typed RUN simultaneously. Both got correct answers.

Kurtz, an early riser who'd gone home to sleep, later admitted he "missed all the fun." The system crashed on average every five minutes during its first month.

But it worked. And Kemeny and Kurtz did something almost unheard of in computing: they gave it away. By 1967, eighty percent of incoming Dartmouth freshmen had learned to program. By the late 1980s, an estimated ten to twelve million schoolchildren had learned BASIC, more people, Kemeny noted, than speak Norwegian.

The consequences reached far beyond education. In January 1975, a young Paul Allen spotted the Altair 8800 microcomputer on the cover of Popular Electronics and rushed to show his friend Bill Gates at Harvard. Gates called the Altair's manufacturer and claimed he had a working BASIC interpreter. He hadn't written a line of code. For eight weeks, Gates and Allen built one from scratch. Allen flew to Albuquerque for the demonstration, typed PRINT 2+2, and the machine printed 4. Microsoft was founded on April 4, 1975. Gates was nineteen.

Within a few years, every major home computer shipped with some version of BASIC burned into its read-only memory. The language

Kemeny and Kurtz built for liberal arts students had become the spark that ignited the personal computer revolution.

Kurtz died on November 12, 2024, at age 96, sixty years and six months after two people typed RUN at 4 AM and both got the right answer.

A SPOTLIGHT ON ERLANG

Nine Nines

Joe Armstrong, a Phone Switch That Could Never Crash, and the Philosophy of Letting Things Fail

In 1972, a young physicist from Bournemouth, England walked out of University College London with a degree in physics and the assumption that he would spend his career studying the natural world. Joseph Leslie Armstrong was smart, curious, methodical, the kind of person drawn to hard problems. But midway through a PhD in physics, funding dried up. Instead of abandoning research entirely, Armstrong did something unusual: he switched fields to computer science. It was a choice that would eventually reshape how the world builds reliable software.

For the next few years, Armstrong worked in artificial intelligence research under Donald Michie, one of the founders of computational intelligence in Britain. He was learning to think like a programmer, to see computation as a design problem rather than a mathematical one. In 1974, he moved to Sweden. In 1984, he joined the Computer Science Laboratory at Ericsson Telecom AB in Stockholm, where he would meet Robert Virding and Mike Williams. The three of them shared a problem that no one could solve: how do you build software that absolutely cannot fail?

The context was brutal. Telephone switches, the massive machines that routed millions of calls across continents, needed to be reachable essentially forever. Ericsson's requirement was 99.9999999 percent uptime. That is nine nines. That is 31 milliseconds of allowable downtime per year. Not per month. Per year. For a machine handling global telecommunications. The question was not academic. Systems that went down cost money. They broke contracts. They damaged lives. A system that crashed for five minutes meant thousands of missed calls, businesses unable to reach customers, emergency services delayed.

The prevailing wisdom said it was impossible. You could write careful code, test exhaustively, design for redundancy, but you could never eliminate failure entirely. The human cost of trying was enormous: armies of engineers debugging, preventing edge cases, building safeguards. And still things broke. Ericsson's own software was written in languages like Prolog, which was declarative and elegant but not designed for the kind of concurrent, distributed systems that phone switches demanded.

Armstrong and his team asked a different question: What if, instead of trying to prevent failure, you designed everything around the assumption that failure would happen? What if you built systems that expected processes to die, that anticipated crashes, that treated errors as information to be acted upon rather than exceptions to be eliminated?

They started by modifying Prolog. They needed concurrency, multiple processes handling multiple calls simultaneously. They needed message passing, so processes could communicate without sharing memory. They needed a way to isolate failures, so one broken process would not corrupt the entire system. Gradually, as they added features and constraints, they realized they were not modifying Prolog anymore. They were building something new. By 1986, they had created Erlang, named after the Danish mathematician Agner Erlang, who had developed traffic theory for telephone exchanges.

The core insight was radical in its simplicity: Stop worrying about how to prevent crashes. Instead, assume things will crash. Build your system so that when something fails, the failure is detected, isolated, and corrected. Make the system self-healing. This became known as the "let it crash" philosophy.

In Erlang, a process is not expensive to spawn or destroy. Processes are lightweight, hundreds of thousands of them can run simultaneously on a single machine, each with its own isolated memory space. When a process encounters an error it cannot handle, it dies. But other processes are watching. They are linked, forming a hierarchy called a supervision tree. A supervisor process sees that a worker process has crashed and simply starts a new one. The system heals itself without human intervention.

This was not defensive programming. This was offensive programming. It rejected the obsessive need to anticipate every possible error. Instead of writing code to handle every exception, you write code that does one thing well, assumes the best case, and lets it crash if something unexpected happens. You write separate code that watches for crashes and fixes them. You build an organization chart for processes: workers do the job, supervisors manage workers, and hierarchies of supervisors manage other supervisors. It mirrors the way human organizations actually work, where it is easier to fire and replace a worker than to prevent all possible mistakes.

Armstrong and his colleagues deployed this philosophy in a system called the AXD301, an asynchronous transfer mode switch that routed data at up to 160 gigabits per second. The AXD301 contained more than a million lines of Erlang. Ericsson announced the switch in March 1998, and in service it reported 99.9999999 percent uptime, extrapolated from a months-long British Telecom trial. The nine nines. Ericsson had built a machine that operated at the edge of human reliability, and it had done so by embracing failure instead of fighting it.

But then something strange happened. In February 1998, Ericsson banned Erlang. The company, in a strategic shift toward becoming a consumer of technology rather than a producer, decided that proprietary languages were liabilities. The ban was a bitter blow. Armstrong and most of the original team had invested years in Erlang, believed in it, proven that it worked. Now their own company was forbidding its use for new projects.

What looked like defeat became liberation. The Erlang team, no longer bound by Ericsson's secrecy, lobbied to open-source the language. On December 8, 1998, Open-Source Erlang was announced to the world. The original developers, discouraged from working at Ericsson, left to start Bluetail AB, and two years later Bluetail was acquired by Alteon WebSystems, itself soon absorbed by Nortel Networks. Erlang, freed from the vault of a single corporation, belonged to everyone now.

Armstrong never stopped working on the problem. In 2003, at the Royal Institute of Technology in Stockholm, he completed a PhD thesis titled "Making reliable distributed systems in the presence of software errors." In that thesis, written after more than a decade of practical experience, he formalized the principles that had emerged from building Erlang: isolation, supervision, message passing, let it crash. The thesis became a foundation for a methodology called OTP, Open Telecom Platform, a set of libraries and design patterns for building robust systems in Erlang.

What no one predicted was how far Erlang's reach would eventually extend. A startup called WhatsApp built its entire messaging platform on top of Erlang. At peak operations, WhatsApp ran billions of users' conversations through servers programmed in Erlang, with a team of roughly fifty engineers maintaining infrastructure that would have required thousands of developers in any other technology. The lightweight processes Armstrong had designed could handle millions of connections per server. The message-passing architecture meant that user messages moved through the system with natural elegance. The supervision trees meant the system could run for years without planned downtime. It worked because it assumed failure from the start.

Discord, the platform that became essential for gaming and communities, is built on Elixir, a language that runs on the Erlang virtual machine. RabbitMQ, the message broker used by countless internet companies, runs on Erlang. A language created to solve a telephone company's problem in the 1980s became the invisible infrastructure of the modern internet.

Armstrong continued teaching and advocating for these principles until his death in April 2019, from pulmonary fibrosis. He remained convinced that bulletproof code was a fantasy, and that the future of computing belonged to systems designed to detect and recover from failure. He believed that once software became complex enough, and it always does, the only honest approach was to assume it would break and build accordingly. In an age of distributed systems, cloud computing, and global networks, that philosophy looks more prescient each year.

The nine nines were not a technical achievement that made Erlang special. They were a signal that a different way of thinking about reliability was possible. A language built on resilience instead of perfection. A system that could never crash because it was designed to crash constantly, in small ways, and recover. This was Armstrong's gift: not a programming language, but a new understanding of what it means to build systems that must never fail.

What happened next

Year	Event
1972	Joe Armstrong graduates from University College London with a physics degree; begins PhD in physics
1974	Armstrong leaves physics, switches to computer science; moves to Sweden
1984	Armstrong joins Ericsson's Computer Science Laboratory; by 1986 develops Erlang with Virding and Williams
1993	Erlang gains internal adoption at Ericsson for telecom applications; OTP framework begins taking shape
1998	AXD301 switch announced; Ericsson bans Erlang in February; Erlang open-sourced December 8
2000	Bluetail AB acquired by Alteon WebSystems (later absorbed by Nortel Networks); Erlang community expands beyond telecom
2003	Armstrong earns PhD with thesis "Making reliable distributed systems in the presence of software errors"
2009	WhatsApp launches, built on Erlang; begins scaling to billions of users
2014	Armstrong becomes professor at Royal Institute of Technology (KTH) in Stockholm

A SPOTLIGHT ON SMALLTALK

"Everything Is an Object"

Alan Kay, Xerox PARC, and the Future That Was Given Away

In 1968, a graduate student at the University of Utah named Alan Kay sketched something the world still hasn't fully built: a portable personal computer for children. He would later call it the Dynabook. It would be cheap enough for schools, powerful enough for simulation, and intuitive enough for a ten-year-old. The hardware to build it didn't exist. Neither did the software. Neither did any institution willing to try.

Kay had an unusual background for a computer scientist. He'd earned a degree in mathematics and molecular biology from the University of Colorado, worked as a professional jazz guitarist, and served as an Air Force programmer. This eclectic foundation gave him a perspective his peers lacked. Where most engineers saw computers as expensive machines for crunching numbers, Kay saw a medium for human expression, as fundamental as the printed book.

In 1970, Xerox opened the Palo Alto Research Center with an extraordinary promise: total intellectual freedom. Xerox, flush with copier profits, wanted to understand the future of information work. They hired the best minds they could find and told them to invent whatever they wanted. What PARC produced in a single decade has few parallels in computing history.

Kay assembled a team called the Learning Research Group. His primary collaborator was Dan Ingalls, a brilliant implementer who turned Kay's philosophical ideas into working systems. Kay conceived the vision; Ingalls built the reality. Their partnership produced Smalltalk, a programming language unlike anything that existed.

In Smalltalk, everything was an object, numbers, strings, classes, even the programming environment itself. There were no special cases. To add two numbers, you sent the message "+" to the first number with the second as an argument. To open a window, you sent a message to a window object. The entire system was built from one consistent idea: objects communicating through messages.

The language was only half the revolution. The other half was what it felt like to use. The entire system was live. You could change code while it was running. You could click on any object and inspect it, modify it, watch the change ripple through the system in real time. There was no compile-wait-crash-restart cycle. You lived inside the system. For children, Kay's original audience, this meant they could build something, watch it break, fix it, and see the result immediately. The computer became something you played, the way a musician plays an instrument. This was Kay's whole philosophy made tangible: the computer as a medium, not a machine.

Smalltalk was only part of what PARC produced. The Alto computer, operational in 1973, was a personal computer more than a decade before the Macintosh. It had a bitmapped display, a mouse, overlapping windows, and was networked via Ethernet, which was also invented at PARC. Laser printing came from PARC too.

In 1979, Xerox's venture capital arm invested in Apple. As part of the deal, Steve Jobs and a team of Apple engineers were given a demonstration at PARC that December. Adele Goldberg, one of the key Smalltalk researchers, initially refused, she understood exactly what Apple would do with what they saw. Xerox management overruled her. What Jobs saw that December set Apple's direction for the next decade. He didn't invent the graphical user interface; he saw it working at PARC and understood, instantly, that this was the future of computing. Apple's Lisa and then the Macintosh were shaped by what Jobs and his engineers witnessed that day. They took no code or hardware with them. In a sense, all they took were ideas. But those ideas would generate more value for others than Xerox had ever created for itself.

The loss was Xerox's. Their management couldn't see the value of what their own researchers had built. Kay left for Atari, then Apple, then Disney. Ingalls continued developing Smalltalk variants for decades; his Squeak environment, started in 1996, became the foundation for Scratch, which has taught millions of children to code, the closest anyone has come to Kay's original Dynabook vision. PARC's inventions generated trillions of dollars of value. Almost none of it went to Xerox.

Smalltalk's influence spread far beyond PARC's walls and far beyond the GUI. Brad Cox combined Smalltalk's message-passing model with C's performance to create Objective-C, which powered Apple's software ecosystem until Swift began replacing it in 2014. Yukihiro Matsumoto cited Smalltalk as a primary influence when he designed Ruby. Smalltalk's prototype-based descendant, Self, directly inspired the object model of JavaScript. The visual IDE that Smalltalk pioneered, where you write, run, and debug in one integrated environment, became the template for modern development environments.

GUIs, object-oriented programming, IDEs, Ethernet, laser printing, the mouse-driven interface, every foundational element of modern personal computing was built at PARC and commercialized by others. As Kay is widely credited with saying, "The best way to predict the future is to invent it." The unspoken irony: Xerox predicted the future and then watched someone else sell it.

In 2003, Kay received the Turing Award for his contributions to object-oriented programming and personal computing. The citation recognized what the industry had been borrowing from for thirty years: the ideas behind Smalltalk, the vision of personal computing, and the conviction that computers should be tools for human thought, not just human labor.

What happened next

Year	Event
1980	Smalltalk-80 publicly released with full documentation and a portable virtual machine
1984	Kay leaves Xerox PARC; joins Atari briefly, then Apple to continue research
1996	Kay and collaborators release Squeak, an open-source Smalltalk; Etoys development begins
2001	Kay founds Viewpoints Research Institute, a nonprofit for children, learning, and software
2003	Kay receives the ACM Turing Award for pioneering object-oriented programming
2008	Pharo forked from Squeak as a streamlined, community-driven Smalltalk
2016	Kay joins HARC at Y Combinator Research to advance computational thinking

Era 3

The Expansion

1980–1989

The 1980s saw object-oriented programming emerge as the dominant paradigm while languages proliferated to serve an expanding universe of computing applications. C++ extended C with classes and inheritance, becoming the language of choice for large systems. Ada, commissioned by the US Department of Defense, set new standards for safety and reliability in critical systems. Smalltalk's influence spread, inspiring Objective-C, which would eventually power Apple's entire ecosystem.

The decade also saw the rise of languages designed for specific domains rather than general-purpose use. MATLAB gave scientists and engineers an interactive environment for numerical computing. Erlang gave Ericsson's telephone switches the fault tolerance to achieve 99.9999999% uptime. Perl gave Unix administrators a Swiss Army knife for text processing. Tcl gave applications an embeddable scripting layer. And Mathematica gave mathematicians a language that understood symbolic computation. By decade's end, the personal computing explosion had made programming relevant to millions of people, and the diversity of languages reflected the diversity of problems computers were being asked to solve.

12 languages in this era

Timeline

- 1980 US Department of Defense approves Ada language for critical systems
- 1983 Bjarne Stroustrup creates C++ at Bell Labs
- 1984 Guy Steele and others publish Common Lisp standard
- 1986 Bertrand Meyer creates Eiffel with Design by Contract; Joe Armstrong creates Erlang at Ericsson
- 1987 Larry Wall creates Perl; Dan Winkler creates HyperTalk for HyperCard
- 1988 John Ousterhout creates Tcl; Stephen Wolfram releases Mathematica
- 1989 Brian Fox creates Bash

World Context

Fall of Berlin Wall (1989); personal computing explosion; IBM PC dominance; rise of GUIs

Dominant Paradigms

Object-oriented programming becomes dominant; functional languages persist in niches; domain-specific languages proliferating; scripting languages emerging

A SPOTLIGHT ON C++

The Thesis That Wouldn't Compile

Bjarne Stroustrup and the Language That Built the Infrastructure

In 1978, a Danish graduate student at Cambridge was trying to finish his PhD, and his programming language kept getting in the way.

Bjarne Stroustrup's thesis required simulating how to distribute operating system services across a network, a problem that demanded modeling computers, processes, and communication channels as distinct objects with distinct behaviors. He chose Simula, the language that had invented object-oriented programming. It was perfect for the task. Classes let him represent each component cleanly. Inheritance let him build hierarchies of behavior. The code was elegant and organized exactly the way the problem demanded.

It was also unbearably slow. Simula's runtime overhead made it impossible to simulate systems at realistic scale. The language was beautiful, but it couldn't do the job.

So Stroustrup rewrote everything in BCPL, a low-level systems language and a precursor to C. Now the simulations ran fast enough. But writing BCPL meant manually laying out bits and bytes to represent the same objects that Simula had modeled so gracefully, no classes, no type checking, no abstraction. The code worked, but the experience was miserable.

Most PhD students would have finished the thesis and moved on. Stroustrup finished the thesis and decided that no programmer should ever have to make that choice again. If you wanted elegance, you shouldn't have to give up speed. If you wanted speed, you shouldn't have to give up elegance. There should be a language that offered both.

He graduated in 1979 under the supervision of David Wheeler, one of computing's pioneers, and immediately joined AT&T Bell Labs in Murray Hill, New Jersey, the building where Dennis Ritchie had created C and where Unix was born. If you were going to build a new systems language on top of C, there was no better place on earth.

By October 1979, just months after arriving, Stroustrup had a working preprocessor called Cpre that added Simula-like classes to C. He called the result "C with Classes." It wasn't a new language yet, it was C with a few features bolted on. But those features were precisely the ones his Cambridge thesis had taught him he needed: classes, derived classes, strong type checking, and inlining. By March 1980, the tool was supporting real projects within Bell Labs, running on sixteen systems.

The language grew slowly, deliberately, over years. Stroustrup added virtual functions for runtime polymorphism. He added operator overloading so programmers could define how mathematical symbols worked with custom types. He built Cfront, a compiler that translated C++ into plain C, a pragmatic trick that made adoption painless, since any machine with a C compiler could now run C++.

In December 1983, a colleague named Rick Mascitti suggested renaming the language. The ++ operator in C increments a value by one. C++: C, improved. Mascitti later said the name was given in a tongue-in-cheek spirit, and he never imagined it would stick. It stuck.

On October 14, 1985, Stroustrup simultaneously released Cfront 1.0 and published The C++ Programming Language. In the absence of a formal standard, the book became the de facto definition of the language. Adoption was immediate. Roughly two thousand programmers used C++ within a year. The number doubled regularly, by 1991, an estimated four hundred thousand programmers were writing C++.

The core of Stroustrup's design philosophy was what he called the zero-overhead principle: you don't pay for what you don't use, and what you do use should be as efficient as anything you could write by hand. A class with no virtual functions should cost nothing more than a C struct. A template should generate code as tight as a manually specialized version. If a feature imposed unavoidable runtime cost, it didn't belong in C++. As he put it: "A facility must not just be useful, it must be affordable in terms of performance."

That principle proved remarkably durable. C++ was standardized by ISO in 1998 and went on to power operating systems, web browsers, game engines, financial trading platforms, and database systems, the invisible infrastructure beneath almost everything. When C++11 arrived in 2011, adding move semantics, lambda expressions, and smart pointers, Stroustrup described it as feeling like a new language. But the zero-overhead principle remained unchanged.

"Within C++, there is a much smaller and cleaner language struggling to get out," he once wrote. He has spent forty-five years trying to let it out, without breaking the code the world already depends on.

What happened next

Year	Event
1985	First commercial release of C++; Stroustrup publishes The C++ Programming Language
1998	C++98 standardized by ISO, the first official international standard
2003	Boost Libraries project gains traction, influencing future standards
2011	C++11 released with move semantics, lambdas, auto, and smart pointers
2017	C++17 adds structured bindings, std::optional, and fold expressions
2020	C++20 introduces concepts, coroutines, modules, and ranges
2024	C++23 published with std::expected, std::print, and deducing this
2026	C++26 finalized by WG21 on March 28

ADA 1980

Jean Ichbiah, U.S. Department of Defense ● Active

By the late 1970s, the US Department of Defense was using over 450 different programming languages across its systems. Maintenance was a nightmare. They held a design competition, and Jean Ichbiah's entry won, a language named after Ada Lovelace, the world's first programmer, built for systems where a bug could kill someone. Ada's strict type system, concurrency features, and formal semantics made it the gold standard for life-critical software.

```
with Ada.Text_IO; use Ada.Text_IO;
procedure Hello is
begin
  Put_Line("Hello, World!");
end Hello;
```

Explicit package imports; Put_Line outputs with newline

Paradigm: Imperative

Typing: Static

Usage: ●●○○○

Lines: 5

★ Ada's formal verification capabilities have proven mission-critical software more reliable than any other language, it's mandated for some aerospace systems.

C++ 1983

Bjarne Stroustrup, Bell Labs ● Active

Bjarne Stroustrup at Bell Labs created C++ as an extension of C, adding object-oriented features while maintaining C's efficiency and close-to-metal control. Stroustrup wanted OOP without sacrificing performance. C++ became the dominant language for building large systems, games, graphics engines, operating systems, where both power and efficiency matter.

```
#include <iostream>
int main()
{
  std::cout << "Hello, World!\n";
}
```

iostream for I/O; std::cout is standard output

Paradigm: Object-Oriented

Typing: Static

Usage: ●●●●○

Lines: 5

★ C++ remained the language of choice for game engines like Unreal Engine and physics simulations, performance-critical code still lived there 40 years later.

COMMON LISP 1984

Guy Steele, Scott Fahlman, others ● Active

By the early 1980s, dozens of incompatible Lisp dialects had fractured the AI research community. Researchers couldn't share code across labs. The standardization effort was legendarily contentious, Guy Steele and Scott Fahlman spent years negotiating between warring factions to forge a common standard. Common Lisp brought together the best ideas from decades of Lisp development, creating an extremely robust language with macros that let programmers extend syntax. It became the language of AI research and remains beloved by passionate programmers.

```
(format t "Hello, World!~%")
```

format t outputs to standard output; ~% newline

Paradigm: Functional

Typing: Dynamic

Usage: ●○○○○

Lines: 1

★ Common Lisp's macro system is so expressive that programmers can define entirely new syntax, it's the closest thing to programming-language-as-abstraction-level.

MATLAB 1984

Cleve Moler, MathWorks ● Active

Cleve Moler was a math professor frustrated that his students spent more time wrestling with FORTRAN than doing actual math. He wrote MATLAB's first version himself in the late 1970s, essentially a wrapper around two linear algebra libraries, and handed it out free on floppy disks. Jack Little saw its potential, convinced Moler to commercialize it, and MathWorks was born. MATLAB became the standard tool for scientific research, used in virtually every engineering discipline.

```
disp('Hello, World!')
```

disp function outputs to command window

Paradigm: Imperative

Typing: Dynamic

Usage: ●●●○○

Lines: 1

★ Moler originally distributed MATLAB free, he didn't think anyone would pay for it. Jack Little proved him wrong, and MathWorks became a billion-dollar company.

OBJECTIVE-C 1984

Brad Cox, Tom Love

● Active

Brad Cox read about Smalltalk's elegant message-passing system and had a radical thought: what if you could graft those ideas directly onto C without giving up C's speed? The resulting hybrid sat in relative obscurity until Steve Jobs chose it for NeXT, and then Apple's acquisition of NeXT made it the language of the iPhone. Objective-C powered Apple development for decades before Swift replaced it.

```
#import <Foundation/Foundation.h>
int main() {
  NSLog(@"Hello, World!");
  return 0;
}
```

@ prefix for string literals; NSLog outputs

Paradigm: Object-Oriented

Typing: Dynamic

Usage: ●●●○○

Lines: 5

★ Steve Jobs chose Objective-C for NeXT in 1988. When Apple acquired NeXT in 1996, Objective-C came with it, an acquisition that accidentally determined the programming language of the iPhone two decades later.

MIRANDA 1985

David Turner, University of Kent ◑ Endangered

David Turner created Miranda as the first widely used lazy functional language, then made a fateful decision: he kept it proprietary. Researchers who wanted an open alternative formed the Haskell committee in 1987, using Miranda as their starting point. Turner's licensing choice accidentally created the most important functional programming language in history.

```
main :: [sys_message]
main = [Stdout "Hello, World!\n"]
```

I/O through message lists; no side effects in the language itself

Paradigm: Functional

Typing: Static

Usage: ●○○○○

Lines: 2

★ Miranda's name comes from the Latin for "to be admired." Turner chose it because he believed the language was beautiful, and by most accounts, he was right.

EIFFEL 1986

Bertrand Meyer, Interactive Software Engineering ● Active

Bertrand Meyer created Eiffel as an OOP language with a philosophy of 'Design by Contract.' Meyer believed that classes should explicitly specify contracts: preconditions, postconditions, and invariants. This makes code more reliable and easier to reason about. Eiffel became influential in formal methods and contract-based programming.

```
class HELLO
create
  make
feature
  make
    do
      io.put_string ("Hello, World!")
      io.put_new_line
    end
end
```

Class with creation procedure, using modern Eiffel syntax

Paradigm: Object-Oriented

Typing: Static

Usage: ●○○○○

Lines: 10

★ Eiffel's 'Design by Contract' reshaped how developers approach software reliability, the idea that contracts can prevent bugs before they happen.

ERLANG 1986

Joe Armstrong, Ericsson ● Active

Joe Armstrong at Ericsson created Erlang for building fault-tolerant telecommunications systems that never stop running. Erlang's concurrency model based on lightweight processes transformed thinking about parallel programming. Systems like WhatsApp, Discord, and AWS use Erlang-derived technology to handle millions of concurrent connections.

```
-module(hello).
-export([start/0]).
start() ->
  io:format("Hello, World!~n", []).
```

Module system with export; io:format outputs

Paradigm: Functional

Typing: Dynamic

Usage: ●●○○○

Lines: 4

★ Erlang's 'let it crash' philosophy turned failure into a feature, instead of preventing crashes, Erlang systems are designed to recover from them automatically.

PERL 1987

Larry Wall ● Active

Larry Wall created Perl as a practical text processing language combining ideas from sed, awk, and shell scripting. Wall believed that programming should be expressive and fun, enabling programmers to write efficient one-liners. Perl became the language of the early web, powering CGI scripts that made the internet interactive.

```
print "Hello, World!\n";
```

Simple print statement with newline escape

Paradigm: Imperative

Typing: Dynamic

Usage: ●●○○○

Lines: 1

★ Perl's motto was 'TMTOWTDI' (There's More Than One Way To Do It), it enabled incredible flexibility but sometimes chaotic code.

HYPERTALK 1987

Dan Winkler ○ Extinct

Dan Winkler created HyperTalk as the scripting language for Bill Atkinson's HyperCard, which shipped free with every Macintosh. Its English-like syntax introduced a generation to programming. Myst, the best-selling PC game of the 1990s, was built in HyperCard. Ward Cunningham's Wiki concept grew from HyperCard, and Brendan Eich's JavaScript both trace their roots to HyperCard and HyperTalk.

```
answer "Hello, World!"
```

answer displays a dialog box with the message and a confirm button

Paradigm: Event-driven

Typing: Dynamic

Usage: ●○○○○

Lines: 1

★ Bill Atkinson later reflected that if he had connected HyperCard stacks over networks, it could have been the first web browser.

MATHEMATICA 1988

Stephen Wolfram, Wolfram Research ● Active

Stephen Wolfram wanted to build a system that could do any computation a mathematician could dream up, symbolic, numerical, graphical, in a single unified language. Wolfram was famously opinionated, and Mathematica reflected his conviction that computation was the fundamental principle of the universe. The result was the most powerful mathematical computing environment ever created, used in virtually every field of scientific research.

```
Print["Hello, World!"]
```

Print function with capital P (Mathematica convention)

Paradigm: Functional

Typing: Dynamic

Usage: ●●○○○

Lines: 1

★ Wolfram later wrote "A New Kind of Science," a 1,280-page book arguing that simple programs underlie all of nature. The book was polarizing, but Mathematica itself was indispensable.

TCL 1988

John Ousterhout, UC Berkeley ● Active

John Ousterhout was watching application after application reinvent its own bad scripting language. He decided to build one embeddable scripting language to rule them all, so simple that its entire syntax fits on a single page. Tcl (Tool Command Language) could be embedded in any application to add scripting capability. Its Tk toolkit made it the fastest way to build graphical interfaces for decades.

```
puts "Hello, World!"
```

puts command outputs to stdout

Paradigm: Imperative

Typing: Dynamic

Usage: ●●○○○

Lines: 1

★ Tcl powered the configuration systems at CERN's Large Hadron Collider, a language simple enough for physicists to use without a CS degree, running the most complex machine ever built.

BASH 1989

Brian Fox, Free Software Foundation ● Active

Brian Fox created Bash (Bourne-Again Shell) as an improved replacement for the Bourne shell, adding history, command-line editing, and more features. Bash became the default shell on Linux systems, used by countless system administrators. It remained compatible with sh while adding modern conveniences.

```
#!/bin/bash
echo "Hello, World!"
```

bash shebang; echo outputs

Paradigm: Imperative

Typing: Dynamic

Usage: ●●●●○

Lines: 2

★ The name Bash stands for 'Bourne-Again Shell', a pun on Stephen Bourne, creator of the original sh. Brian Fox wrote Bash in 1989 as a free replacement for sh.

Era 4

The Internet Age

1990–1999

The 1990s transformed computing through the internet, and programming languages evolved to meet the demands of web development. Python and Java rose to prominence during this era, each addressing different needs: Python for education, scripting, and rapid prototyping; Java for enterprise systems with its promise of write-once-run-anywhere portability. JavaScript, created in just ten days by Brendan Eich, emerged as the lingua franca of web browsers despite its hasty design.

The internet itself drove language innovation in ways no one predicted. PHP, created to track web page hits, became the server-side language powering billions of pages. Ruby combined Perl's practicality with Smalltalk's elegance, and would later spark the Rails revolution. Scripting languages made it trivial to write small, concise programs without compilation. Haskell unified the academic world's lazy functional languages into a single rigorous standard. By 1999, the landscape was clear: multiple languages would coexist, each optimized for different domains and communities. The idea of a universal language had died, replaced by a rich ecosystem of specialized tools.

13 languages in this era

Timeline

- 1990 — International committee creates Haskell to unify lazy functional languages
- 1991 — Guido van Rossum creates Python; Alan Cooper and Microsoft release Visual Basic
- 1993 — Lua created by PUC-Rio team in Brazil; R created by Ihaka and Gentleman in New Zealand
- 1995 — Yukihiro Matsumoto creates Ruby; James Gosling releases Java; Brendan Eich creates JavaScript in 10 days; Rasmus Lerdorf creates PHP
- 1996 — Xavier Leroy and team release OCaml
- 1998 — Macromedia releases ActionScript for Flash

World Context

World Wide Web goes mainstream; dot-com boom; rise of internet culture; Netscape vs. Internet Explorer

Dominant Paradigms

Multiple paradigms coexist; scripting languages ascendant; Java brings OOP to enterprise; JavaScript dominates browsers; functional programming finds a home in Haskell and OCaml

A SPOTLIGHT ON PYTHON

The Christmas Project

Guido van Rossum and the Hobby Language That Conquered the World

---

Over Christmas 1989, Guido van Rossum was bored. He was a thirty-three-year-old programmer at the Centrum Wiskunde & Informatica in Amsterdam, the office was closed for the holidays, and he needed a project.

He had spent the previous four years working on ABC, a programming language designed to be clean, readable, and accessible to non-programmers. ABC had gotten a lot of things right. It used indentation to define code blocks instead of braces or keywords, an idea most programmers at the time considered bizarre. It had powerful built-in data types. Its syntax read almost like English. But ABC had failed commercially, and van Rossum understood why: it was a closed system. You couldn't extend it. You couldn't call C libraries from it. You couldn't adapt it to solve problems its designers hadn't anticipated. ABC was elegant but sealed shut.

Van Rossum also had a practical problem. He was working on the Amoeba distributed operating system at CWI, and the team needed a scripting language for system administration, something between C and the Unix shell. Nothing on the market fit.

So he started writing a new language. He kept ABC's good ideas, the indentation-based syntax, the readable structure, the powerful built-in types, and threw out everything that had made it rigid. The new language would be extensible. It would integrate easily with C. It would have a large standard library, because van Rossum believed a language should come with batteries included. And it would prioritize readability above almost everything else, because code is read far more often than it is written.

He named it Python, after Monty Python's Flying Circus. The documentation from those early days is threaded with references to dead parrots and spam.

On February 20, 1991, van Rossum posted Python 0.9.1 to the alt.sources newsgroup, twenty-one uuencoded parts that anyone with a Unix machine could download and compile. It already had classes, exception handling, functions, and the core data types (lists, dictionaries, strings) that would define the language. The reception was small but enthusiastic. A community began forming almost immediately.

What happened over the next decade was steady, unglamorous growth. Python found niches in system administration, text processing, web development, and scientific computing. Van Rossum moved to the United States in 1995, working first at the Corporation for National Research Initiatives, then briefly at a startup, before joining Google in 2005. At Google, he created Mondrian, an internal code review tool written in Python. The company gave him the freedom to spend half his time on Python itself.

The community gave him a title: Benevolent Dictator For Life. Van Rossum made the final call on language design decisions, and for nearly two decades, the arrangement worked. Python grew methodically, adding generators, decorators, list comprehensions, and a gradually expanding standard library, without the design-by-committee drift that had fragmented other languages.

Then the world changed around it.

In 2012, a University of Toronto research team used a neural network to win the ImageNet competition by a historic margin, and the deep learning revolution began. The researchers had used Python. When Google released TensorFlow in November 2015 and Facebook released PyTorch in January 2017, both chose Python as their primary interface. Python was not fast, and everyone knew it. But its readability and flexibility made it the fastest language to think in. Scientists could prototype an idea in an afternoon, iterate for a week, and hand the performance-critical pieces to C or CUDA underneath. Python became the front end of the most consequential technology shift since the internet.

By the 2020s, Python was the world's most popular programming language by virtually every measure. It dominated artificial intelligence, data science, web development, automation, and education. The Christmas hobby project of a bored Dutch programmer had become the language in which the future was being written.

Van Rossum almost didn't see it through. In July 2018, exhausted by a contentious debate over a minor syntax feature, the "walrus operator," PEP 572, he resigned as BDFL. "I don't ever want to have to fight so hard for a PEP and find that so many people despise my decisions," he wrote. The community transitioned to a steering council model. Van Rossum retired in 2019, un-retired in 2020 to join Microsoft, and as of 2025 was still working on making Python faster.

His design philosophy was distilled by Tim Peters into the Zen of Python, a set of aphorisms that became the language's soul. The most famous: "There should be one (and preferably only one) obvious way to do it." It was the opposite of Perl's motto, the opposite of C++'s multi-paradigm flexibility, and the opposite of the chaos that had characterized programming for decades. Van Rossum bet that clarity would win. He was right.

What happened next

Year	Event
1994	Python 1.0 released with lambda, map, filter, and reduce
2000	Python 2.0 released with list comprehensions and Unicode support
2005	Django released; NumPy 1.0 unifies the scientific computing ecosystem
2008	Python 3.0 released, intentionally breaking backward compatibility
2015	TensorFlow released by Google; Python becomes the default for ML
2017	PyTorch released by Facebook AI Research; AI dominance accelerates
2018	Van Rossum resigns as BDFL; community transitions to steering council
2020	Van Rossum retires, then joins Microsoft to work on performance
2021	Python reaches #1 on TIOBE index, surpassing Java and C

HASKELL 1990

International committee (Peyton Jones, Hughes, Wadler, others) ● Active

An international committee of functional programming researchers created Haskell to unify competing lazy functional languages. Named after logician Haskell Curry, Haskell became the reference language for functional programming research. Its pure functional model, lazy evaluation, and flexible type system attracted academics and inspired the design of modern languages.

```
main = putStrLn "Hello, World!"
```

putStrLn outputs with newline

Paradigm: Functional

Typing: Static with Inference

Usage: ●●○○○

Lines: 1

★ Haskell's purity and laziness enable reasoning about programs mathematically, you can prove correctness like a mathematical proof, not just test.

A SPOTLIGHT ON HASKELL

The Supercooled Solution

Simon Peyton Jones, No PhD, and the Committee That Built Something Beautiful by Accident

In the fall of 1987, a young lecturer from University College London stopped at Yale on his way to a conference in Portland, Oregon. Simon Peyton Jones was twenty-nine years old. He did not have a PhD. He did not have a research group. What he had was a single-minded obsession with functional programming and a gift, already evident to anyone who had heard him talk, for making difficult ideas feel like the most exciting thing in the world. A colleague would later say that Peyton Jones could read the telephone directory and make it sound thrilling. At Yale, he wanted to see Paul Hudak, an associate professor five years older, warm and modest, the sort of host who would insist on carrying your bag to the guest room and then stay up half the night talking about lambda calculus. Philip Wadler, another functional programming researcher, also stopped at Yale and endorsed the same conclusion. The problem was this: there were too many lazy functional languages.

By the mid-1980s, more than a dozen of them existed, all built around the same radical idea. These languages were lazy. They would not evaluate an expression until its result was actually needed. They were pure. Calling a function with the same input would always produce the same output, no hidden side effects, no surprises. The people who worked on them were captivated. The simplicity was addictive, the connection to pure mathematics intoxicating. It was, as several of them would later write, like a drug. But at Yale, at Chalmers in Sweden, at Oxford, at Nijmegen, everywhere, researchers were designing their own lazy languages and busily implementing them. David Turner's Miranda was the most prominent, running at 120 sites, but there were at least a dozen others, all roughly the same aside from the syntax. They had built a Tower of Babel.

Peyton Jones had arrived at this community by an unusual route. He was born in 1958 at a naval base near Cape Town, South Africa, where his father was stationed as an officer in the Royal Navy. The family returned to England when he was a few months old, moved to Trinidad when he was three, and back to England when he was eight. At boarding school, he encountered his first computer, an IBM machine with a hundred memory locations, programmed in pure machine code. There was no way to save your work. Every session, you typed the whole program in from scratch. At Trinity College, Cambridge, he enrolled to read mathematics. The Trinity mathematicians were, as he later put it, "streets ahead" of him. He transferred to electrical sciences, graduated with first-class honors, and completed the Cambridge Diploma in Computer Science. It was, and would remain, his only formal qualification in the field. In a world where the doctorate is the price of admission to academic life, Peyton Jones simply walked in without one.

He got a lectureship at UCL on the strength of one publication. With no famous adviser and no research group, he sought out David Turner, the leading functional programmer in Britain, and met him every few months for coffee and guidance. At UCL, Peyton Jones was doing pioneering work on compiling lazy functional programs, showing that laziness, which everyone assumed would be slow, could be compiled into efficient code. Turner's Miranda was the language the whole community orbited around. And it was Turner who would unknowingly set the course of Peyton Jones's entire career.

At the Portland conference, Peyton Jones and Hudak announced an impromptu meeting to discuss creating a single, common functional language. The situation, as they would later describe it, was like a supercooled solution, an unstable state that needed only a single random event to precipitate crystallization. The committee's first instinct was practical: adopt Miranda and evolve it. After a brief and cordial exchange, Turner said no. He was committed to a single language standard and did not want multiple dialects in circulation. He asked them to make their new language sufficiently distinct from Miranda. He also declined to join the committee.

Turner's refusal meant they had to design everything from scratch. Designing from scratch was harder and would take longer. But it gave them freedom to contemplate radical approaches, freedom to make choices Miranda would never have made. If they had started from Miranda, it seems unlikely they would have developed the features that became Haskell's most distinctive contributions to computer science.

The first meeting took place at Yale in January 1988, hosted by Hudak. They needed a name. Wadler suggested a process: names on a blackboard, then each person crosses out what they dislike. The board filled up. Semla, Vivaldi, Mozart, CFL, Curry, Haskell B Curry, two dozen others. One name survived: Curry, after the mathematician Haskell Brooks Curry. That night, two members realized the puns would be relentless. The spice. The concept of currying. And the horrifying thought: Tim Curry, since TIM was Jon Fairbairn's abstract machine, and Tim Curry was famous for The Rocky Horror Picture Show. The next day they settled on "Haskell." Hudak later visited Curry's widow, Virginia, to ask permission. Her parting remark: "You know, Haskell actually never liked the name Haskell."

The committee that Peyton Jones had helped convene began pulling in two directions. They wanted a language that was simple and elegant. They also wanted one that was useful. Richard Bird, a distinguished computer scientist at Oxford, resigned in mid-1988, warning of "a severe danger that the principles of simplicity, ease of proof, and elegance will be overthrown." Tony Hoare, in a memorable letter, offered a wistful prediction: Haskell was "probably doomed to succeed." Peyton Jones's own instinct, "it has to be simpler," the principle he would describe as guiding most of his research life, shaped the committee's choices. They embraced what they called superficial complexity while eschewing deep complexity. They avoided extensible records and parameterized modules. And they adopted type classes, a feature that complicated everything but was too good to miss.

Type classes were born from a misunderstanding. Wadler conceived of them after a conversation with Joe Fasel following one of the Haskell meetings. Fasel had a different idea in mind, but he had the key insight that overloading should be reflected in the type of a function. Wadler misunderstood what Fasel meant, and from that misunderstanding conceived type classes, a more radical and far-reaching mechanism than either man had imagined. Wadler's student Steven Blott helped formalize the type rules and proved the system sound. They were adopted by acclamation. No one anticipated that type classes would transform Haskell from a common language into a laboratory for ideas about types that had never been tried before. It was a happy coincidence of timing and a happy accident of miscommunication.

There was another problem that became a triumph. Haskell's commitment to purity meant a function could have no side effects, no reading files, no printing to screens. In practice, this made basic input and output painfully clumsy. For years the committee tried various approaches. Then, in 1989, Eugenio Moggi published a paper using an idea from category theory, monads, to describe features of programming languages. Wadler was at Glasgow with Peyton Jones. He read Moggi's work and saw something no one else had: the technique Moggi had used to structure mathematical semantics could be turned directly into a programming construct. He wrote a paper called "Comprehending monads." But Wadler's insight was general, not aimed at any particular problem. It was Peyton Jones and others at Glasgow who, reading Wadler's paper, felt the pieces click into place. Monads could solve the embarrassment: they provided an ideal framework for sequencing side effects inside a pure language without compromising its mathematical guarantees. In retrospect, the designers came to believe that laziness mattered less for its own sake than for what it forced on them: because the language was lazy, it had to be pure, and because it was pure, they had to invent monadic I/O. It was, as Peyton Jones later put it, deeply satisfying to see an apparently esoteric bit of mathematical logic become so directly useful to solve a practical problem.

The Haskell Report was published on April 1, 1990. The date was mostly an accident, but close enough to keep. A year or two later, Hudak sent an April Fool's email saying he was quitting Yale for a music career. David Wise immediately phoned to plead with him to reconsider. In 1998, John Peterson wrote a bogus press release claiming Microsoft had adopted Haskell. Not long after, Peyton Jones announced his real move from Glasgow to Microsoft Research in Cambridge, an event Peterson knew nothing about at the time.

Peyton Jones would spend twenty-three years at Microsoft Research, a corporate laboratory that let him pursue fundamental work without commercial pressure. He became the principal architect of the Glasgow Haskell Compiler, the implementation that made Haskell real. After the committee disbanded in 1999, he took on sole editorship, his role evolving into what he called Benign Dictator of Linguistic Minutiae. He was, by all accounts, the sort of person who would spend an hour helping a stuck colleague untangle a problem that was not his own, and who met confusion by admitting that he found it confusing too.

Paul Hudak, who had hosted the first meeting, visited Virginia Curry in her home, and once told a room full of language designers that the best standard was one everyone complained about equally, died of leukemia on April 29, 2015, at sixty-two. Wadler, whose misunderstanding had produced Haskell's most influential feature, became a professor at Edinburgh. Both Peyton Jones and Wadler were elected Fellows of the Royal Society. In 2022, Peyton Jones received an OBE for services to education and computer science. He had spent years co-founding the Computing at School initiative that transformed the British national curriculum. He left Microsoft to join Epic Games, working on a new language called Verse.

The language he helped build went on to influence nearly every modern programming language. Facebook rebuilt its spam-fighting infrastructure in Haskell. Features it pioneered (type classes, monadic I/O, lazy evaluation, higher-kinded polymorphism) appeared in Rust, Scala, Swift, and Kotlin. The Haskell community remained small enough and agile enough that it usually not only absorbed language changes but positively welcomed them. As the designers put it, it was like throwing red meat to hyenas.

A language born from a supercooled solution, crystallized by a chance visit to

Yale, forced into originality by a licensing refusal, named after a man who never liked his own name, published on April Fool's Day, distinguished by features that arose from a misunderstanding, and embarrassed into one of its greatest innovations by the inability to print to a screen. A committee language that should have been a compromise and instead became something people called beautiful. Held together for decades by a man who never did get that PhD, though he has since collected four honorary doctorates. Probably doomed to succeed.

What happened next

Year	Event
1990	Haskell 1.0 defined with lazy evaluation, algebraic data types, and type classes
1999	Haskell 98 standard published, providing a stable, portable specification
2002	GHC becomes the dominant implementation under Peyton Jones's leadership
2015	Facebook deploys Sigma, a Haskell spam filter handling 1M+ requests per second
2020	Haskell Foundation established to support the ecosystem and tools
2021	GHC 9.0 released with simplified subsumption system
2022	Peyton Jones receives an OBE for services to education and computer science
2024	Peyton Jones joins Epic Games to develop Verse for the Unreal Engine

PYTHON 1991

Guido van Rossum ● Active

Guido van Rossum was bored during his Christmas holidays in 1989 and wanted a language that felt like reading pseudocode. He named it after Monty Python, not the snake. Van Rossum's insistence on readability , enforced through significant whitespace, created a language so clear that it became the accidental lingua franca of AI and data science. Python is now the most taught, most used, and most versatile programming language on Earth.

```
def main():
    print("Hello, World!")
main()
```

def main() defines the entry point; main() must be called to execute

Paradigm: Multi-paradigm

Typing: Dynamic

Usage: ▼▼▼▼▼

Lines: 3

★ Python's dominance in AI and data science is so complete that machine learning researchers expect their tools to be Python-first; it shaped modern AI.

VISUAL BASIC 1991

Alan Cooper

◐ Endangered

Alan Cooper at Microsoft created Visual Basic as a visual programming environment extending BASIC. Cooper's key insight was that graphical UI building should be as easy as writing code. Visual Basic made Windows programming accessible to hobbyists and small businesses, generating an enormous ecosystem.

```
MsgBox "Hello, World!"
```

MsgBox displays message dialog

Paradigm: Imperative

Typing: Dynamic

Usage: ●●○○○

Lines: 1

★ Visual Basic created an entire generation of accidental programmers, people who never intended to code but found they could build useful applications.

A SPOTLIGHT ON POWERSHELL

Which Part of Windows Is Confusing You?

Jeffrey Snover, a Demotion, and the Command Line That Saved Microsoft's Cloud

The engineer sat across from his executives, pitch-ready, eager. Jeffrey Snover had spent more than a year on his vision: a command-line shell that could make Windows as scriptable and automatable as Unix had been for decades. He believed he understood what Windows desperately needed. He was wrong about how his company felt about it.

"Exactly which part of fucking Windows is confusing you?" they asked him, more than once, in the halls where promotions were decided and product roadmaps were drawn. The question wasn't seeking clarification. It was a dismissal, a refusal, a cultural wall that said: we are a GUI company. Bill Gates had declared cmd.exe Windows's final command line. That was settled. Done. No more command-line tools. GUIs were the future. GUIs were where the money was.

Snover did not accept this.

To understand the depth of his conviction, you need to understand the scale of what he was swimming against. Microsoft in the late 1990s and early 2000s was not a company that believed in command-line interfaces. It was a company that had designed the entire soul of Windows around graphical windows, pull-down menus, clickable buttons, visual feedback. The Unix world, meanwhile, was humming. Linux administrators could string together commands, pipe outputs, automate whole data centers with scripts. They could manage systems without leaving the keyboard. Windows administrators had to click. Click, click, click. For every task, a GUI dialog. For every variation, a new dialog sequence. It did not scale.

Snover saw this gap. He saw it clearly. And in 2002, he put pen to paper and wrote what became known as the Monad Manifesto, a white paper that laid out, with technical precision and philosophical clarity, what Windows system administration could become if it embraced the power of the command line. The manifesto described the "Monad" (eventually called PowerShell): an object-oriented shell that would pipe actual .NET objects, rich with properties and methods that downstream commands could inspect and manipulate without error-prone text parsing, instead of the mere text that limited Unix pipes. The vision was elegant. The vision was technically sound. The vision was also, in Microsoft's eyes, seditious. What followed were years that Snover himself would later describe as "the most miserable two or three years of my life." He was a senior architect. He had standing. He had permission, ostensibly, to pursue interesting projects. And then, when he refused to abandon PowerShell, when he kept working on it, kept refining it, kept believing in it even as the organization turned cold, Microsoft made the choice that would define his relationship with the company for the next two decades. They demoted him. Significantly. The financial penalty was real. The career hit was public enough that Snover, out of what he would later call "profound embarrassment," told no one about it for over a decade. Not his colleagues. Not his peers. Only his wife.

The thing about Snover, though, was that he had already made a decision: the title mattered less to him than the work. He had reframed the demotion, in his own mind, as the price of admission. "I've got a certain amount of time," he would later explain, "and I want the world to be different and hopefully better because I existed." That was the operating principle. Not promotions. Not favor. Not the comfortable climb up the ladder. The world, better. So he kept going. Through the wall of resistance, through the organizational hostility, through what felt like everyone saying no. His small team kept building. They refined the object pipeline. They implemented cmdlets, lightweight commands that would be discoverable and composable. They built the execution engine. They wrote documentation that nobody asked for, to a standard that exceeded what anyone would have expected.

By November 2006, PowerShell version 1.0 was released. It came out first for Windows XP Service Pack 2, Windows Server 2003 Service Pack 1, and Windows Vista. Cautiously. Carefully. Without the institutional push that other Microsoft products received. The reception was... enthusiastic, among those who understood scripting. Within half a year, it had been downloaded nearly a million times. People got it. Administrators felt the relief in their bones. But the vindication, the real vindication, came later.

Azure happened. Cloud computing at scale required something that the mouse could never deliver: automation. Repeatability. The ability to provision a hundred machines, configure them identically, deploy an application across them, scale up, scale down, all without human fingers touching keyboards. Cloud is fundamentally a problem of orchestration. You cannot click your way through a cloud infrastructure. You need a command line, scripting, piping. You need the thing that Snover had fought for so hard, and that Microsoft had tried to strangle in its crib. When Azure launched and began to grow, suddenly PowerShell was not a nice-to-have administrative quirk. It became essential infrastructure. The architects building Azure realized that without PowerShell, without a way to automate, to script, to pipe objects through commands, the entire cloud vision became impractical. PowerShell, the tool the organization had demoted its architect for inventing, would enable Microsoft's transformation into a cloud company. The thing they had wanted to kill saved them.

By 2009, three years after PowerShell's release, Snover was promoted to Distinguished Engineer, the highest individual contributor rank at Microsoft. No team to manage. No budget to oversee. Just the recognition that he had been right, and the organization had been wrong, and that the intellectual integrity to refuse a demotion and keep building anyway had earned him a place in the company's most elite technical ranks.

PowerShell went open-source in 2016. It became cross-platform. Today it runs on Linux and macOS alongside Windows. It is no longer a Windows tool; it is the language of infrastructure automation across operating systems, used by cloud engineers, by DevOps practitioners, by every organization that has embraced the principle that administration at scale requires code, not clicks. Snover's vision of the object pipeline became so normal, so obvious, that people forget it was ever controversial.

The lesson Snover carried forward was unsparing: "Be willing to take the demotion or face the criticism to keep your eye on the long-term prize." Not every battle is won by convincing the room. Sometimes it is won by refusing to stop working, even when the room would prefer you did. Sometimes it is won by choosing what you believe is right over what you are praised for. Sometimes the cost is real, professional embarrassment, financial penalty, years of underappreciation. But if you are right about what the world needs, and if you are willing to pay that cost, the vindication eventually comes. Not always quickly. But it comes.

Snover left Microsoft in 2022, after more than two decades, and joined Google as a Distinguished Engineer in site reliability engineering. He eventually retired from Google and now spends his time as what he calls a "Philosopher-Errant," attending conferences and giving talks about systems safety and risk management. The man who fought to give Windows a command line now advises on the engineering practices that keep infrastructure safe. The work continues.

The principle never changed: make the world different and better because you existed.

What happened next

Year	Event
2002	Jeffrey Snover writes the Monad Manifesto, outlining the vision for a command-line shell for Windows
2003	PowerShell (then called Monad) development begins; Snover is demoted for refusing to abandon the project
2006	PowerShell 1.0 released in November for Windows XP SP2, Server 2003 SP1, and Vista; downloaded nearly 1 million times in six months
2009	Snover promoted to Distinguished Engineer at Microsoft, the organization's highest individual contributor rank
2016	PowerShell goes open-source and becomes cross-platform (PowerShell Core), running on Linux and macOS
2018	Snover becomes Chief Architect for the Azure Storage and Cloud Edge group as Azure scales globally
2022	Snover leaves Microsoft after 23 years; joins Google as a Distinguished Engineer in Site Reliability Engineering
2026	Snover retires from Google to pursue philosophical and technical writing on systems safety and engineering culture

DYLAN 1992

Apple Advanced Technology Group ◐ Endangered

Apple's Advanced Technology Group created Dylan (Dynamic Language) as a next-generation language combining the power of Lisp with a more conventional syntax. Apple envisioned it as the future of Macintosh programming, but corporate upheaval in the mid-1990s killed the project. Dylan was open-sourced and lives on through a small but dedicated community, a ghost of Apple's most ambitious programming language bet.

```
module: hello
format-out("Hello, World!\n")
```

Lisp semantics with Algol-like syntax; module system is first-class

Paradigm: Multi-paradigm

Typing: Dynamic

Usage: ●○○○○

Lines: 2

★ Dylan was originally called 'Ralph.' Apple renamed it, but the project's internal codename persisted in documentation for years. Even Apple's naming instincts aren't always perfect on the first try.

LUA

1993

Ierusalimschy, Celes, de Figueiredo, PUC-Rio

● Active

Roberto Ierusalimschy and colleagues at PUC-Rio in Brazil created Lua as a lightweight, embeddable scripting language. Lua was designed to be minimal but powerful, embedding easily in C/C++ applications. It became the standard scripting language for games; World of Warcraft, Roblox, and countless others used Lua.

```
print("Hello, World!")
```

print() outputs text; Lua compiles to bytecode for embedding in C programs

Paradigm: Imperative

Typing: Dynamic

Usage: ●●●○○

Lines: 1

★ Lua's ability to embed in games made it the foundation of modding communities, players worldwide use Lua to extend their favorite games.

A SPOTLIGHT ON CLOJURE

The Savings Account

Rich Hickey, Two Years Without a Paycheck, and the Case Against Mutation

In the late 1990s, Rich Hickey worked as a software consultant in the financial technology sector, building high-performance trading systems in C++, Java, and C#. He was good at it, the work was lucrative, intellectually demanding, and had the peculiar intensity that comes with managing millions of dollars in code. But something gnawed at him. Every large system he built accumulated the same category of bugs, the kind that emerge only under load, in production, when real money is moving. Concurrency bugs. Race conditions. Data races. The kind of problems that happen when multiple threads reach for the same piece of mutable state at the same time, and there is no certainty about the order in which they will arrive. He watched teams spend weeks debugging issues that should have been impossible, because the code looked correct on every thread's individual journey. The problem was the convergence. The problem was change.

By 2004, Hickey was certain of the root cause. Programming languages treated shared mutable state as the default, the natural way to represent reality. You created an object, and you modified it. This seemed intuitive, it mirrored how we think about the physical world, and it was deeply wrong. The moment a second actor could reach the same mutable data, the system became a graph of dependencies that no human mind could fully hold. Locking helped, but locking was a band-aid on a conceptual error. What if you simply didn't mutate things? What if data was immutable by default, and you built new values instead of modifying old ones?

He started sketching a programming language in his head. Not a theoretical exercise. A real language that could run real programs. But to build it, he would need time. Uninterrupted time. Time without the pressure of billable hours and client deliverables. In 2005, at the height of the dot-com recovery, when consulting fees were at their peak, Rich Hickey made a decision that his financial advisor probably did not recommend: he cashed out his retirement savings and took a sabbatical to build a programming language that nobody had asked for, on a platform nobody wanted, based on ideas everyone said were impractical. He liquidated approximately two and a half years' worth of living expenses. He set up alone in front of a machine and began to write. The language he was building was Clojure, a new dialect of Lisp that would run on the Java Virtual Machine. Lisp was unpopular, a language from the 1950s that had never escaped academia and specialist circles. The JVM was for Java, a language designed to be explicit and verbose and corporate. The combination seemed perverse. But Hickey had an insight: immutable data structures and the JVM's garbage collector were made for each other. If you couldn't modify data, you could share versions you'd left behind. And Lisp, austere, flexible, homoiconic Lisp, was the ideal vehicle for expressing these ideas. There would be no syntax noise, no declarations, no ceremony. Just functions and data.

For two and a half years, he worked. Sometimes alone in a room for weeks at a time. No venture capital, no angel investors, no promise of commercial viability. He did ninety percent of the implementation himself, from the compiler to the standard library. He released Clojure to the world in October 2007 as an open-source project. For a year afterward, he continued to work unpaid, refining the language. By 2008, people were starting to notice. Not many, but the right people: programmers who recognized that Hickey was articulating something true about why software got so complicated.

In 2009, Clojure reached version 1.0, and Rich Hickey released a statement that would echo through the community for years: "I have borne the costs of developing Clojure myself, but 2009 is the last year I, or my family, can bear that." His retirement account was gone. The savings account that was meant to fund his later years had been spent on a programming language that, at that moment, barely anyone used. The economy was in freefall. He had a family to support. The mathematics were clear and ruthless: he could continue Clojure on an empty bank account, or he could find consulting work and let the language slowly atrophy. Neither option was acceptable.

The community responded. Within a day of his plea for funding, more than one hundred individuals and five companies had pledged money. The target was exceeded. Clojure would live.

What made people respond? The technical merits were real, but they were not the whole story. Hickey's talks had begun to articulate a philosophy that resonated with programmers across languages. In 2009, he delivered "Are We There Yet?", a talk that took aim at object-oriented programming as it was actually practiced, not the theory, but the thing you did in your IDE every day. He argued that objects pretended to model reality but actually modeled change, bundling identity and state together in ways that made systems impossible to reason about. In 2010, he gave a talk simply titled "Clojure, Made Simple," and in 2011, he stood on a stage at Strange Loop and delivered "Simple Made Easy," a talk that would define a generation's thinking about software complexity. In that talk, he drew a distinction most programmers had never thought to make. Simple did not mean easy. Easy meant accessible, familiar, convenient. Simple meant the opposite of complex, it meant lacking in components braided together, lacking in dependencies and surprising connections. A system could be easy to build and terribly complex. And a system could be simple, but require discipline to use. Most programming languages, he argued, chose the easy path, which meant they accumulated complexity. Clojure chose simplicity: immutable data structures, no side effects by default, a small set of core abstractions. This made it harder to use at first, there was no familiar "class," no comfortable patterns to reach for. But it made systems simpler in the end.

His word for what made things complex was "complecting", braiding things together that should stay separate. Identity from state. Code from data. Mutable containers from the values they held. Every time you complected these things, you made your system harder to reason about, and every concurrent program broke because of complecting. Clojure was an attempt to uncomplect programming, to separate concerns that had never been separated before.

By 2012, major companies were building systems in Clojure. Walmart used it for price optimization. Nubank, a digital bank in Brazil that would eventually become the world's largest, built its core infrastructure on Clojure. For the first time, the people who worked with Hickey's language were having a visceral, productive experience that matched the theory. Systems that were simpler. Bugs that simply didn't exist because the data couldn't be mutated out from under you.

The personal sacrifice of those early years was never forgotten. There was no dormitory project here, no venture capital in a hoodie pocket, no garage startup with sweat equity and options that might be worth something someday. This was a middle-aged consultant, already successful, walking away from a lucrative career and betting his family's financial security on an idea. He could have been wrong. The idea could have failed. Clojure could have remained a curiosity, a Lisp dialect that nobody used, and Rich Hickey could have spent the rest of his career rebuilding wealth that he had deliberately burned. That he was right, and that the language survived, was the result of both intellectual clarity and personal courage.

The savings account was empty. But it had purchased something worth far more: a new way of thinking about how software should be built, and a community of programmers who would spend the next fifteen years teaching others the principles Hickey had learned, alone in his room, over two and a half years of unpaid work.

What happened next

Year	Event
2005	Rich Hickey liquidates retirement savings and begins Clojure development
2007	Clojure 0.1 released in October as open-source project
2008	Hickey continues unpaid development, building ecosystem and refining design
2009	Clojure 1.0 released in May; Hickey's funding plea raises community support in December
2010	Rich Hickey delivers influential talks; Metadata Partners founded to develop Datomic
2011	"Simple Made Easy" talk at Strange Loop becomes paradigm-defining address on software complexity
2012	Cognitect formed when Metadata Partners merges with Relevance consultancy; Datomic released
2014	Nubank adopts Clojure as primary language for core financial systems
2020	Nubank acquires Cognitect, solidifying Clojure's position in production banking

R

1993

Ross Ihaka, Robert Gentleman, University of Auckland

● Active

Ross Ihaka and Robert Gentleman at University of Auckland created R as an open-source implementation of the S statistical language. R became the language of statistical computing, enabling researchers and data scientists to analyze data and create visualizations.

```
cat("Hello, World!\n")
```

cat outputs without the vector formatting that print() adds

Paradigm: Functional

Typing: Dynamic

Usage: ●●●○○

Lines: 1

★ R is so dominant in statistics that academic papers without R code are viewed with suspicion, it's the lingua franca of statistical research.

DELPHI/OBJECT PASCAL 1995

Borland ◐ Endangered

Borland created Delphi as an Object-Oriented extension of Pascal with a full-featured visual IDE. Delphi showed that Pascal could shine in serious Windows programming. Developers loved it for its clean syntax and full-featured IDE, though it never achieved Java or C++'s ubiquity.

```
program HelloWorld;
begin
  WriteLn('Hello, World!');
end.
```

WriteLn outputs with newline; no imports needed

Paradigm: Object-Oriented

Typing: Static

Usage: ●○○○○

Lines: 4

★ Delphi's visual IDE was so effective that developers could build applications faster than Java competitors; it proved design matters.

JAVA 1995

James Gosling, Sun Microsystems ● Active

James Gosling at Sun Microsystems created Java with the vision 'Write Once, Run Anywhere.' Java's JVM (Java Virtual Machine) ran on any platform, liberating programs from hardware lock-in. Java became the enterprise language, powering banking systems, e-commerce platforms, and large-scale applications worldwide.

```
public class HelloWorld {
  public static void main(String[] args) {
    System.out.println("Hello, World!");
  }
}
```

Class-based; System.out.println outputs

Paradigm: Object-Oriented

Typing: Static

Usage: ●●●●●

Lines: 5

★ Java's 'Write Once, Run Anywhere' vision transformed computing; the JVM became so successful it now runs Kotlin, Scala, Clojure, and other languages.

A SPOTLIGHT ON JAVA

The Set-Top Box That Ate the World

James Gosling, a Failed TV Project, and the Language That Became the Infrastructure

---

In December 1990, three programmers at Sun Microsystems started a skunkworks project with a peculiar mission: build the software that would run inside televisions, cable boxes, and toasters. The company believed consumer electronics were about to become programmable, and they wanted to own the platform. James Gosling, a thirty-five-year-old Canadian who had spent six years at Sun building windowing systems, was put in charge of the language. The team would eventually grow to thirteen.

Gosling had grown up in a suburb of Calgary, a self-described "ridiculously geeky kid" who discovered computers at thirteen by wandering into the University of Calgary and dumpster-diving for discarded punch cards. The cards contained code fragments and sometimes passwords. He taught himself to program by reassembling what other people had thrown away. By high school, he was writing software to analyze satellite data for the university's physics department.

He earned his PhD at Carnegie Mellon, where he wrote Gosling Emacs, an influential text editor whose source code contained an ASCII skull and crossbones above a particularly difficult algorithm, a warning to future programmers not to touch it. He also built a PERQ emulator that translated code written for one machine into instructions another machine could execute. It was a small project. He did not yet know he was prototyping the idea that would define Java.

After a brief, miserable stint at IBM, which he later called "a teaching moment about how cool tech never wins out over stupid bureaucracy," Gosling joined Sun Microsystems in 1984. He built NeWS, a windowing system based on PostScript that was technically superior to the competing X11 standard. It lost anyway, because Sun kept it proprietary while X11 was open. The lesson was painful and formative: the best technology doesn't always win. Openness matters.

The Green Project, as Sun called the television effort, was supposed to produce the software for a new generation of smart consumer devices. Gosling started by trying to extend C++, the dominant systems language. He gave up. C++ was too dangerous for consumer electronics. A single bad pointer could crash a set-top box, and you couldn't ask a family watching television to reboot their cable box and recompile. He needed something safer.

So he wrote a new language. He called it Oak, after the tree outside his office window. Oak looked like C++ but stripped out everything that could hurt you: no pointer arithmetic, no manual memory management, no multiple inheritance. He added garbage collection, strong type checking, and a virtual machine that could run the same compiled code on any hardware. Its promise was "Write Once, Run Anywhere." It was the PERQ emulator idea, scaled up to an entire language.

By the summer of 1992, the team had a working prototype: the Star7, a handheld touchscreen device with an animated interface. A graphic artist named Joe Palrang created an animated character, a cartwheeling software agent, to guide users through the interface. They called it Duke. The Star7 was a remarkable piece of engineering: a touchscreen smart device built four years before the PalmPilot and fifteen years before the iPhone.

Nobody wanted it. The interactive television market that Sun had bet on did not materialize. Cable companies were not interested in smart set-top boxes. The Star7 was too expensive, too far ahead of its time. By early 1994, the Green Project was dying. Gosling's language existed, elegant and homeless, looking for a problem worth solving.

Then the World Wide Web happened.

One of the team members built a web browser in Oak, a clone of Mosaic that could run small programs downloaded from the internet. They called it WebRunner, after the movie *Blade Runner* . It was later renamed HotJava. The browser was the proof of concept the project had been missing. Here was Oak's real purpose: not television, but the internet. A language whose code ran identically on every machine was exactly what a network connecting millions of different machines needed.

In early 1995, the team discovered that "Oak" was already trademarked by a hardware company. They needed a new name. During a brainstorming session, someone threw out "Java." The name came from the Indonesian island famous for its coffee, and the team had consumed enough of it during late-night coding sessions that the reference felt earned. It stuck.

On May 23, 1995, at the SunWorld conference, Sun and Netscape announced that Java would be built into Netscape Navigator, the browser used by roughly eighty percent of the internet. Within months, Java applets, small programs running inside web pages, were everywhere. Dancing animations, stock tickers, chat rooms, interactive demos. For a brief, dizzying period, Java was the language of the web.

The applets didn't last. But Java found something more durable: the enterprise. Banks, airlines, insurance companies, and governments adopted Java for their back-end systems. Its safety features, the same ones Gosling had designed to protect a television viewer from a crashed set-top box, turned out to be exactly what you wanted when processing millions of financial transactions. The verbosity that programmers mocked, the five lines required just to print Hello World, was the point. Every keyword was a guardrail. The language designed for your living room ended up running your paycheck, your flight reservation, and your tax return.

Then came the wars. Microsoft licensed Java in 1996 and built Visual J++, a version that added proprietary extensions tying programs to Windows. It shipped with the "Java Compatible" logo while failing Sun's own compatibility tests. Sun sued in October 1997. A federal judge barred Microsoft from the Java trademark and forced compliance with Sun's standards. By January 2001, Microsoft had paid twenty million dollars and was frozen at Java 1.1.4, prohibited from further development. The court order killed one product and gave birth to another. The story of what came next belongs to C#.

In 2007, Sun open-sourced Java as OpenJDK. It was the lesson from NeWS, applied twenty years late: the best technology doesn't win unless people can own it. The move probably saved the language. It meant that when Oracle later acquired Sun, no single company could lock Java away.

But Java's success had already created its own prison. "Within a couple of years of being launched, it became really clear that we couldn't change anything that would break anyone's code," Gosling later reflected. Billions of lines of Java ran in production systems worldwide. Every improvement had to be backward-compatible with code written in 1996. The JVM, meanwhile, outgrew its creator: Scala, Clojure, Kotlin, and Groovy all chose to run on Java's virtual machine rather than build their own. The platform Gosling had designed to run one language now ran dozens. Android, built on a Java-compatible runtime, put it in three billion pockets. The virtual machine had become more important than the language itself.

In January 2010, Oracle acquired Sun Microsystems. Gosling stayed for four months. He had been demoted, his bonuses eliminated, his authority over Java, his own creation, removed. He later described sitting in integration meetings and watching an Oracle lawyer's eyes "sparkle" when the discussion turned to patent leverage against Google. He quit on April 2, 2010. His blog post was cryptic: "Just about anything I could say that would be accurate and honest would do more harm than good."

He drifted to Google briefly, then to Liquid Robotics, where he wrote software for autonomous ocean drones, then to Amazon Web Services as a Distinguished Engineer. He retired in July 2024. In interviews, he kept returning to the same self-description: "I'll always be an engineer. When I go to the grave, I'll still be an engineer."

The kid from Calgary who taught himself to program from garbage had built a language for televisions that nobody watched, which became the language of a web that hadn't been invented yet, which became the infrastructure of a global economy that couldn't function without it. Java's Hello World, with its public class, its public static void main, its System.out.println, is the most ceremonious greeting in all of programming. Five lines just to say hello. Gosling would say that's the point. Every one of those words is there to protect you from something.

JAVASCRIPT 1995

Brendan Eich, Netscape ● Active

Brendan Eich at Netscape created JavaScript in 10 days as a scripting language for web browsers. Originally called Mocha, then LiveScript, it was renamed JavaScript in a marketing deal. JavaScript was designed to be easy to use for HTML authors, not expert programmers. Despite its rushed creation, JavaScript became the language of the web.

```
console.log('Hello, World!');
```

console.log outputs to browser console

Paradigm: Multi-paradigm

Typing: Dynamic

Usage: ●●●●●

Lines: 1

★ JavaScript's 'Good Parts' are brilliantly designed, while its 'Bad Parts' (like type coercion) bedevil developers, yet it dominates the web.

PHP 1995

Rasmus Lerdorf ● Active

Rasmus Lerdorf created PHP as a set of scripts to track visitors to his online resume, then released them publicly. Other developers kept adding features, and PHP grew into a language almost by accident, no grand design, just relentless pragmatic accretion. It powered the early web because it was easy to deploy: just drop a .php file on a server and it worked. PHP now powers over 75% of server-side web, including WordPress, Facebook (before Hack), and Wikipedia.

```
<?php echo "Hello, World!"; ?>
```

PHP tags required; echo outputs

Paradigm: Imperative

Typing: Dynamic

Usage: ●●●●○

Lines: 1

★ PHP powers over 75% of all websites with known server-side programming language, despite its critics, PHP's accessibility created the modern web.

RUBY 1995

Yukihiro Matsumoto ● Active

Yukihiro Matsumoto created Ruby in Japan, inspired by Perl, Smalltalk, and Python. Matsumoto believed programming should be fun and expressive, prioritizing programmer happiness. Ruby's elegance and flexibility made it beloved by developers, especially through the Ruby on Rails web framework that democratized web development.

```
puts "Hello, World!"
```

puts outputs with newline

Paradigm: Multi-paradigm

Typing: Dynamic

Usage: ●●●○○

Lines: 1

★ Ruby on Rails made web development so accessible that hobbyists could build serious applications, Twitter, GitHub, Shopify, and Basecamp all started as Rails apps.

OCAML 1996

Xavier Leroy, Inria ● Active

Xavier Leroy at Inria created OCaml (Objective Caml), extending the ML language with object-oriented features. OCaml combined functional programming, pattern matching, and OOP in a cohesive design. It became the favorite language for formal verification and compiler development.

```
let () = print_endline "Hello, World!"
```

let binding; print_endline outputs with newline

Paradigm: Multi-paradigm

Typing: Static with Inference

Usage: ●●○○○

Lines: 1

★ OCaml is used to build theorem provers and formal verification systems; its type system catches bugs other languages miss.

ACTIONSCRIPT 1998

Macromedia ◐ Endangered

Before YouTube, before the iPhone, Flash was how you experienced multimedia on the internet. ActionScript powered that entire world, from Homestar Runner to experimental art installations to the earliest streaming video players. Gary Grossman at Macromedia built ActionScript as an ECMAScript dialect to give Flash developers a real programming language. When Steve Jobs refused to put Flash on the iPhone in 2010, he signed ActionScript's death warrant.

```
trace("Hello, World!");
```

trace outputs to debug console

Paradigm: Object-Oriented

Typing: Dynamic

Usage: ●●○○○

Lines: 1

★ ActionScript powered the entire Flash ecosystem, web animations, games, and rich interfaces from 1996 until HTML5 replaced it.

A SPOTLIGHT ON PERL

The Linguist's Language

Larry Wall, Perl, and the Duct Tape That Held the Internet Together

Larry Wall grew up wanting to be a missionary. Born in 1954 in Los Angeles, he studied at Seattle Pacific University, where he earned a degree in Natural and Artificial Languages, a program that straddled linguistics and computer science. He went on to graduate work in linguistics at UC Berkeley and planned to join the Summer Institute of Linguistics, where he would travel to remote communities and create writing systems for languages that had never been written down. Health problems that made fieldwork impractical ended that plan. Wall never made it overseas. Instead, he applied the instincts of a linguist to a very different kind of language.

In 1987, Wall was working at Unisys as a system administrator. He needed to generate reports from a stream of text files spread across a network, configuration management data that had to be extracted, transformed, and formatted. The standard Unix tools could handle pieces of the job: sed for text substitution, awk for pattern-based processing, shell scripts for gluing it all together. But none of them could do the whole thing. The problem fell in the gap between tools that were too simple and languages that were too complex.

So Wall wrote his own. He released Perl 1.0 on December 18, 1987, through the comp.sources.misc Usenet newsgroup. The original name was "Pearl", after the parable of the pearl of great price, but when Wall discovered an existing language with that name, he dropped the a. Then he backronymed it: Practical Extraction and Report Language. (He also offered an alternative expansion: Pathologically Eclectic Rubbish Lister.) What made Perl different from every other language of its era was that Wall had designed it the way a linguist designs a language, not the way a computer scientist does. He believed programming languages should work like natural languages, messy, context-dependent, redundant, and flexible. In English, you can say the same thing a dozen different ways, and the listener figures out what you mean from context. Wall wanted that for code. He codified this as Perl's motto: TIMTOWTDI, There's More Than One Way To Do It, pronounced "Tim Toady." It was a deliberate philosophical stance against the idea that a programming language should enforce a single correct approach. Wall argued that different programmers think differently, and a language that accommodates that diversity will be more useful than one that demands conformity.

In a 1998 interview, Wall explained the philosophy with a story about the

University of California at Irvine: when the campus was first built, they put in buildings but no sidewalks, just grass. The next year, they came back and paved the sidewalks where the trails were. "Perl is that kind of a language," Wall said. "It is not designed from first principles. Perl is those sidewalks in the grass." He also defined what he called the three virtues of a great programmer: laziness (writing code that saves effort in the long run), impatience (building tools that anticipate what the user needs next), and hubris (writing code so clean you wouldn't be ashamed to show it). It was funny, but he meant it.

Then the World Wide Web happened, and Perl was in exactly the right place at exactly the right time.

The early web ran on CGI, the Common Gateway Interface, a protocol that let a web server execute a program and return the result as a web page. CGI scripts needed to do exactly what Perl was built for: parse text input, process data, and generate formatted output. By the mid-1990s, Perl was the dominant language for web development. Amazon's early infrastructure used Perl. Craigslist ran on it. Slashdot ran on it. IMDB was built with it. The language was everywhere, and it became known as "the duct tape that holds the internet together."

In October 1995, the Comprehensive Perl Archive Network (CPAN) launched, creating the first major open-source module repository. Before package managers became universal, before npm or pip or Cargo existed, CPAN showed what a community-maintained library ecosystem could look like. At its peak, it hosted over 200,000 modules.

Perl also found an unlikely second home in bioinformatics. The Human Genome Project, which sequenced the entire human genetic code, relied heavily on Perl for sequence analysis. DNA is, after all, a string, and string processing was what Perl did better than anything else alive.

The decline, when it came, was gradual. PHP offered simpler web development for people who didn't need Perl's power. Python offered cleaner syntax for people who found Perl's flexibility overwhelming. Ruby, inspired partly by Perl's expressiveness, attracted developers who wanted elegance alongside convenience. Perl didn't collapse; it receded. The duct tape was still there, but the internet had moved on to newer adhesives.

Wall announced Perl 6 in July 2000, promising a ground-up redesign. It would take fifteen years to reach production readiness and would eventually be renamed Raku in 2019, an acknowledgment that it had become a different language entirely, not a successor. Wall had been slowed by health issues and by the sheer ambition of the project. In the meantime, Perl 5 carried on, maintained by a dedicated community that still keeps it among the top languages in active use. The missionary linguist never made it to the field. But his conviction that languages should be human-shaped, flexible, forgiving, and rich with ways to say the same thing, produced a tool that carried the early internet on its back and proved that the best programming languages aren't designed like engineering blueprints. They're designed like living languages, grown from the patterns of actual use.

A SPOTLIGHT ON JAVASCRIPT

JavaScript in 10 Days

In May 1995, Netscape Communications was at war with Microsoft for control of the web browser. Marc Andreessen wanted a scripting language embedded directly in the browser, something lightweight that designers could use to make web pages interactive. He recruited Brendan Eich, a 33-year-old programmer who had joined Netscape expecting to work on Scheme.

Instead, Andreessen told Eich to build a new language. It had to look like Java (for marketing), behave like Scheme (for flexibility), and be ready for Netscape Navigator 2.0's beta release. Eich had 10 days.

Working from May 6 to May 15, Eich created Mocha, later renamed LiveScript, then JavaScript. He borrowed first-class functions from Scheme, prototypal inheritance from Self, and a C-like syntax. The result was messy, full of quirks, and shipped with known bugs. None of this mattered. JavaScript shipped in the browser, and the browser was everywhere. By the time developers noticed the flaws, millions of web pages depended on the language. There was no going back.

What happened next was swift. ECMAScript was standardized in 1997; Microsoft reverse-engineered the language for Internet Explorer. By 2005, Ajax had made JavaScript capable of real applications. Google's V8 engine, released in 2008, made it fast enough for serious computation. Node.js arrived in 2009 and let JavaScript escape the browser entirely, running on servers for the first time. TypeScript added static types in 2012. ES6 overhauled the language in 2015 with classes, modules, and arrow functions. By 2026, JavaScript remained the most-used programming language in the world.

Brendan Eich had ten days, a set of contradictory requirements, and no time to get it right. The language he shipped was messy, full of quirks, and embedded in every browser on earth before anyone could fix it. Three decades later, it runs everywhere, and the ten-day deadline is still the most famous origin story in programming.

Era 5

The Oddballs

1972–2018

This chapter breaks from chronology to collect programming's most playful and provocative experiments, languages designed not to be useful, but to ask what "useful" even means. Some, like Brainfuck, strip programming down to its bare mathematical minimum. Others, like Shakespeare, ask whether code can also be literature. Together, they reveal deep truths about the nature of computation while refusing to take themselves seriously.

Every esoteric language in this collection is a thought experiment wearing a disguise. INTERCAL satirized the pomposity of language design. Malbolge asked whether a language could be deliberately impossible. LOLCODE proved that internet culture and computer science could collide productively. And Rockstar demonstrated that a programming language could double as a power ballad. We return to the timeline with the Modern Era on the next page.

7 languages in this era

A SPOTLIGHT ON VISUAL BASIC

The Dropout and the Demo

Alan Cooper, a Counterculture Kid, and the Tool That Made Windows Programmable

---

In the summer of 1985, Alan Cooper found himself watching a demonstration that would redirect the entire trajectory of his career and, without knowing it yet, shape how millions of programmers would work for the next two decades. He stood before a Xerox Star workstation, an expensive, futuristic machine that felt to him like a glimpse into a better world. The GUI was fluid. The interface was intuitive. It made Windows feel like something a cave person had designed with a stick and a rock. In that moment, Cooper had an epiphany: someone needed to build a tool that would let ordinary people create graphical applications as easily as the Star made it look.

Cooper was not an obvious choice to change the world of software. He had dropped out of high school in Marin County, California, a rebellious kid drawn to the Bay Area's counterculture in the early 1970s. He drifted between interests before landing on architecture, studying it at College of Marin while simultaneously learning COBOL, an odd combination that somehow worked. When the personal computer revolution sparked to life, Cooper saw an opportunity. He bought an IMSAI 8080 and, in 1976, founded Structured Systems Group with a radical idea: create serious business software for machines people actually owned, not just corporations. SSG became one of the earliest PC software companies, proving that a pair of ambitious programmers could compete with the industry giants.

But Cooper had learned something early that would take the rest of the industry two decades to understand: what he considered easy to understand and use was not what normal computer users considered easy. This realization drove him to found Access Software in 1980, his second startup, dedicated to designing software that users would find easy to learn and use. At Access, Cooper pioneered two concepts new to the PC world. The first was separating the user-facing design process from the computer-facing design process, an idea that would not see widespread recognition for another twenty years. The second was blind-testing usability during development, not after the software was finished. Usability testing existed in academia, but Cooper was doing it in real time, redesigning interactions on the fly based on what he observed. Access created and sold a spreadsheet to a small publisher called Systems Plus, but the arrival of the IBM PC hit the industry like a bomb and wiped out hundreds of small startups, including Systems Plus. Lacking customers, Access paid its bills and closed its doors.

Cooper kept building. Between Access and what came next, he invented and sold two more products to publishers, both successful. Each release sharpened his understanding of how software should feel in a user's hands. By 1985, Cooper was independent again, working on his own vision. What he wanted to build was not a programming tool. It was a shell construction set, a tool that would let users, generally corporate users, create a bespoke shell for their own particular needs. He called his prototype Tripod, and it was going to be visual. A user would drag a control onto a form, arrange the interface to match their workflow, and the shell would respond. Everything else, the plumbing, the message handling, the boring infrastructure, would vanish.

Tripod was never going to be pretty. Cooper describes the prototype as "about 15,000 lines of ugly C code." But it worked. It showed, with undeniable clarity, what was possible. The controls were responsive. The development flow was fast. The forms you built looked like real Windows applications.

Cooper showed the Tripod prototype to several software publishers before bringing it to Microsoft. In the spring of 1988, he demonstrated it to Bill Gates and several of his associates. He presented it as what it was: a shell construction set. Cooper expected Tripod to ship with Windows 3 as its shell. Gates agreed generally, though in retrospect he was not fully committed to that plan. Because other publishers had already seen the prototype, Cooper informed Gates that he would rename the product as soon as they signed a deal. He chose "Ruby," which would have been his second child's name if Marty had been born a girl.

Cooper spent the next year building a new version of Tripod from scratch, one that could pass Microsoft's rigid quality assurance program. The "15,000 lines of ugly C code" became more than 30,000 lines of rock solid code, built by Cooper's three-person team.

But here is where the story takes an unexpected turn. The OS/2 group at Microsoft objected to Ruby. They wanted Windows 3 to have a shell identical to theirs. Despite the fact that one could replicate the OS/2 shell in a few minutes with Ruby, the battle was lost. Ruby became an orphan inside Microsoft.

Then, after some months and without Cooper's knowledge, Gates made a decision that would redirect the trajectory of the entire Windows ecosystem. He chose to use Ruby as a new front end for QuickBASIC, a language that had been languishing in Microsoft's product lineup. QuickBASIC was powerful enough to do serious work, but nobody loved it. Nobody was excited about it.

It was utilitarian. Boring. Exactly the kind of thing that Windows, with its colorful buttons and popup windows, seemed to be rendering obsolete. Because Cooper's code was modular, Microsoft's developers could detach his shell language and connect BASIC in its place. Visual Basic was born.

Visual Basic 1.0 shipped in May of 1991, and what happened next was an earthquake in the industry that nobody had quite anticipated. Suddenly Windows programming was accessible to people who had never written graphical code before. A designer could become a programmer. A COBOL developer stuck in mainframe purgatory could suddenly build applications that looked modern and felt responsive. You dragged controls, you wrote event handlers, and your application worked. No hunting through Microsoft's tangled C libraries. No wrestling with the Windows API. No existential despair.

Cooper's influence on Visual Basic was profound, though his name was systematically erased from the marketing. Microsoft's middle managers, moving fast and loose with the truth, claimed the product as their own invention. Only later, when the industry's historians began paying attention, did they give Cooper the recognition he had negotiated into his contract. In 1994, Bill Gates himself presented Cooper with the first Windows Pioneer Award. The first book dedicated entirely to Visual Basic would call him the "Father of Visual Basic," though Cooper himself knew the relationship between his vision and what Microsoft built was complicated.

The irony of Alan Cooper's career is not subtle, but it is not the irony most people assume. Cooper's intentions were always to create easy-to-use software for users. From SSG to Access Software to the products he sold between them, his obsession was the person sitting in front of the screen, not the person writing the code behind it. Tripod was designed for users. Ruby was designed for users. But what emerged from Microsoft's internal politics and Gates's pragmatic redirection was something Cooper never intended: easy-to-use software for programmers. For a generation of developers in the 1990s, Visual Basic would be their first experience writing real software. Some of them never moved beyond it. Others used it as a foundation to learn deeper languages and concepts. All of them owed something to a kid from Marin County who had dropped out of high school, spent fifteen years obsessing over how to make software easier for users, and accidentally created the tool that made programming easier instead.

The lessons Cooper learned across twelve product releases at SSG, one at Access, and three as an independent author gave him the confidence to start Cooper Interaction Design in 1992, the first interaction design consulting company. Three years later, all of that knowledge was published in the first edition of About Face, which would establish him as the father of modern interaction design. He had spent decades building for users. His most famous programming tool was built for users too. It just ended up in the hands of programmers.

What happened next

Year	Event
1976	Alan Cooper founds Structured Systems Group (SSG), creating the first serious business software for microcomputers
1980	Cooper founds Access Software, pioneering blind usability testing during development and user-centered design for PC software
1985	Cooper witnesses Xerox Star workstation demonstration and conceives of Tripod as a shell construction set for users
1988	Cooper demonstrates Tripod to Bill Gates in the spring; Microsoft purchases it; Cooper renames it Ruby
1989	OS/2 group at Microsoft blocks Ruby as the Windows 3 shell; Ruby becomes an orphan inside the company
1990	Gates decides to pair Ruby with QuickBASIC without Cooper's knowledge, creating Visual Basic
1991	Visual Basic 1.0 officially released in May; becomes an immediate success
1992	Cooper founds Cooper Interaction Design, the first interaction design consulting company
1994	Bill Gates awards Cooper the first Windows Pioneer Award for his pioneering work
1995	Cooper publishes About Face, establishing himself as a leading voice in interaction design
1997	Visual Basic becomes the most popular programming language among Windows developers
2002	VB.NET released, replacing classic Visual Basic with a .NET-based successor

Timeline

- 1972 — Don Woods and James Lyon create INTERCAL as a parody of 1970s languages
- 1993 — Urban Müller creates Brainfuck in under 1KB; Chris Pressey creates Befunge on a 2D grid
- 1998 — Ben Olmstead creates Malbolge, designed to be impossible to program
- 2001 — Karl Hasselström and Jon Åslund create Shakespeare Programming Language
- 2007 — Adam Lindsay creates LOLCODE in internet cat-speak
- 2018 — Dylan Beattie creates Rockstar, programs that read as power ballad lyrics

World Context

Spanning five decades of computing culture, from mainframe humor to internet memes to rock and roll

Dominant Paradigms

Turing completeness as art form; minimalism (Brainfuck), hostility (Malbolge), literary play (Shakespeare, Rockstar), and internet culture (LOLCODE)

INTERCAL 1972

Don Woods, James Lyon

ESOTERIC ● Active

Don Woods and James Lyon created INTERCAL as a parody of programming languages, deliberately designed to be difficult and absurd. INTERCAL's statements include 'FORGET', 'RESUME', and 'COME FROM', intentional jokes mocking programming language features. Despite (or because of) its absurdity, INTERCAL became a beloved cult language with a devoted community.

```
DO ,1 <- #13
PLEASE DO ,1 SUB #1 <- #238
DO ,1 SUB #2 <- #108
DO ,1 SUB #3 <- #112
DO ,1 SUB #4 <- #0
DO ,1 SUB #5 <- #64
DO ,1 SUB #6 <- #194
DO ,1 SUB #7 <- #48
PLEASE DO ,1 SUB #8 <- #22
DO ,1 SUB #9 <- #248
DO ,1 SUB #10 <- #168
DO ,1 SUB #11 <- #24
DO ,1 SUB #12 <- #16
DO ,1 SUB #13 <- #162
PLEASE READ OUT ,1
PLEASE GIVE UP
```

Each number encodes a character via bit-reversal; READ OUT decodes and prints the array

Paradigm: Esoteric

Typing: Dynamic

Usage: ●○○○○

Lines: 16

★ INTERCAL's most infamous feature is COME FROM, which jumps backwards, the opposite of GOTO, a joke that actually revealed something true about goto's problems.

BEFUNGE

1993

Chris Pressey

ESOTERIC ● Active

Chris Pressey created Befunge as an esoteric language where the code itself is a 2D grid through which the instruction pointer moves like a spaceship. Befunge challenged conventional thinking about program flow, creating a playful language where code is spatial and visual.

```
"!dlroW ,olleH">:#,_@
```

2D grid; " pushes string as ASCII, >:#,_ loops to print each character, @ ends

Paradigm: Esoteric

Typing: Untyped

Usage: ●○○○○

Lines: 1

★ Befunge showed that programming could be spatial and visual, creating a whole category of 2D languages that shaped generative art.

BRAINFUCK 1993

Urban Müller

ESOTERIC ● Active

Urban Müller created Brainfuck as a joke, the smallest possible Turing-complete language with only 8 commands. Despite its absurdity, Brainfuck became a beloved esoteric language, spawning a community of enthusiasts who enjoy the challenge of writing useful programs in an intentionally torturous language.

```
++++++++[>++++[>++>+++>+++>+<<<<-]
>+>+>->>+[<]<-]>>.>---.+++++++
..+++.>>.<-.<.+++.------.
--------.>>+.>++.
```

106 characters, 8 commands; outputs the classic form "Hello World!"

Paradigm: Esoteric

Typing: Untyped

Usage: ●○○○○

Lines: 4

★ Brainfuck's existence demonstrates that Turing-completeness requires almost nothing, you can compute anything with 8 operations, though it's absolute torture.

MALBOLGE 1998

Ben Olmstead

ESOTERIC ● Active

Ben Olmstead created Malbolge as an esoteric language so difficult that it took months to write even a 'Hello, World!' program. Malbolge uses ternary logic, encrypted operations, and self-modifying code, making it one of the most challenging languages ever created.

```
('&%:9]!~}|z2Vxwv-,POqponl$Hjig%eB@@>}=<M:9wv6WsU2T|nm-,jcL(I&%$#"
`CB]V?Tx<uVtT`Rpo3NlF.Jh++FdbCBA@?]!~|4XzyTT43Qsqq(Lnmkj"Fhg${z@>
```

Encrypted operations; no readable syntax

Paradigm: Esoteric

Typing: Untyped

Usage: ●○○○○

Lines: 1

★ It took the Malbolge community 2 years to write the first nontrivial program, it's a language where achievement is measured in days of work per line.

SHAKESPEARE PROGRAMMING LANGUAGE 2001

Karl Hasselström, Jon Åslund

ESOTERIC ● Active

Karl Hasselström and Jon Åslund created Shakespeare as an esoteric language where code looks like a Shakespearean play. Variables are characters, operations are dialogue, it's one of the most creative esoteric languages, blending programming with literature.

```
The Infamous Hello World Program.
Romeo, a young man.
Juliet, a likewise young woman.
Act I: The Only Act.
Scene I: The One Scene.
[Enter Romeo and Juliet]
Juliet: Open your heart!
[Exeunt]
```

Truncated; full program requires 80+ lines of Shakespearean dialogue to encode ASCII values

Paradigm: Esoteric

Typing: Dynamic

Usage: ●○○○○

Lines: 8

★ The full Shakespeare Hello World program is over 80 lines long. Each character's 'worth' is computed from adjectives and nouns in dialogue, 'a fine gentle flower' sets a variable to ((2×2)+1)×2×2.

LOLCODE 2007

Adam Lindsay ESOTERIC ● Active

Adam Lindsay created LOLCODE as an esoteric language based on internet cat meme language ('lolspeak'). LOLCODE programs look like funny cat conversations, yet they're Turing-complete. It's a humorous language that attracted a community of meme-loving programmers.

```
HAI 1.2
  CAN HAS STDIO?
  VISIBLE "Hello, World!"
KTHXBYE
```

Lolspeak syntax; HAI starts, VISIBLE outputs

Paradigm: Esoteric

Typing: Dynamic

Usage: ●○○○○

Lines: 4

★ LOLCODE has a complete specification with conditionals (O RLY? / YA RLY / NO WAI), loops (IM IN YR LOOP), and even type casting (MAEK). Someone built a web server in it.

ROCKSTAR 2018

Dylan Beattie

ESOTERIC ● Active

Dylan Beattie created Rockstar as an esoteric language where code reads like 1980s rock song lyrics. Variables are declared with lines like 'Tommy is a rebel', loops use 'while', and output uses 'scream'. It's Turing-complete yet reads like poetry.

```
Shout "Hello, World!"
```

Shout outputs; poetry-like syntax

Paradigm: Esoteric

Typing: Dynamic

Usage: ●○○○○

Lines: 1

★ Beattie created Rockstar after noticing job postings asking for 'rockstar developers.' In Rockstar, FizzBuzz reads like a breakup ballad, complete with verses, choruses, and heartbreak.

A SPOTLIGHT ON RUBY

The Language That Wished You Well

Yukihiro Matsumoto, Programmer Happiness, and a Name Chosen Before the First Line of Code

On February 24, 1993, in an online chat that would ripple across two decades of computing, Yukihiro Matsumoto and Keiju Ishitsuka made a choice. They were designing a new programming language from the ground up, thinking through its philosophy, its purpose, its possibilities. They knew they wanted to name it after a precious jewel. Gemstones seemed fitting for something meant to gleam in the programmer's hand. They considered "Coral," which had a nice ring, elegant and oceanic. Matsumoto chose "Ruby" instead: shorter, lovelier, and carrying a whisper of the language it would eventually challenge, Perl, that other jewel in the programmer's toolkit. None of this code existed yet. They had not written a single function or a single class definition. They named it before they built it. They decided who it would be before it was born.

That is how Ruby began, and the beginning is unusual for this book. Most languages emerge from a crisis or an urgent problem crying out for solution. Ruby came from dissatisfaction: the kind that lingers in a programmer's mind as they write code in languages that almost work, almost feel right, but never quite do. Matsumoto, born April 14, 1965, in Osaka Prefecture, had been teaching himself programming since childhood; by high school he had moved on to teaching himself whole languages. By the time he reached the University of Tsukuba in the early 1980s, studying information science under Ikuo Nakata in the programming languages and compilers laboratory, he had a clear eye for what languages did well and where they failed.

He saw Perl and was dazzled but uneasy. Larry Wall's creation was powerful, brilliant even, at text manipulation, but Matsumoto found it clever and baroque, designed in the end for a narrow purpose. The philosophy behind it was "there's more than one way to do it," and Matsumoto loved that freedom, even admired it. He wanted something more balanced, though. He looked at Python and saw a language that claimed object-oriented design but made it feel like an afterthought, a bolt-on feature rather than the foundation. This bothered him deeply. Object orientation, as Matsumoto understood it, was a way of thinking, a way of organizing the world, and if you believed that, it had to run through the entire system like blood through veins. Bolting it on missed the point.

What Matsumoto wanted was impossible: a language powerful enough for real work, practical for scripting, object-oriented to its marrow rather than by add-on, and beautiful. A language that made the act of programming feel less like fighting the machine and more like thinking aloud. In his own words, he was trying "to make Ruby natural, not simple." The distinction matters. Simple is easy to understand quickly. Natural is something deeper. Natural means the language follows the patterns of how humans think and work.

So in 1993, shortly after that chat where the name was chosen, Matsumoto began writing Ruby. It took him months just to get basic functionality working. In August of that year, he managed to write the line that produced "Hello world," that tiny phrase that marks a language's entry into existence. He synthesized what he loved from his many influences: the expressiveness of Perl, the purity of Smalltalk, the rigor of Eiffel, elements of Ada, and the elegance of Lisp. He was building something that had never existed before: a language that cared about making you happy.

For nearly three years, Ruby lived in obscurity. Matsumoto refined it, added features, talked about it on Japanese online forums where early adopters began to tinker. Then on December 21, 1995, he released Ruby 0.95 to the world. Version 0.95 was incomplete and rough in places, but it had what he had built it to have: real object-orientation from top to bottom, elegant syntax that read almost like poetry, and a philosophy that put the programmer's experience at the center. Within days, three more versions followed, each one getting closer to a vision that was becoming clearer.

By 1997, the small community around Ruby was growing. Matsumoto was hired by Netlab, a Japanese software consulting group, to work on Ruby full-time. Consider how extraordinary that was: at a moment when programming languages were the domain of academics and corporate research labs, a company had bet on this young Japanese programmer's belief that what the world needed was a language designed around programmer happiness.

What drove this commitment was a philosophy that grew around Matsumoto's own character. He was known for patience, for kindness, for taking criticism seriously and responding thoughtfully. When trolls appeared on Ruby mailing lists in the early days, when some of the usual internet rudeness took hold, community members began invoking a simple acronym: MINASWAN. Matz Is Nice And So We Are Nice. The acronym set the community's terms, and it announced where the language's culture came from: directly from its creator's temperament. You could feel Matsumoto's character in the way Ruby behaved, in its flexibility, in the assumption that you were a thoughtful programmer who deserved a tool that trusted you.

Ruby should have remained a language of the Japanese programmer community, respected but obscure. It grew popular in Japan, yes: by the year 2000, only five years after its release, Ruby was more widely used than Python there. But the rest of the world barely noticed. Then in 2005, something changed. A Danish programmer named David Heinemeier Hansson, who had been using Ruby to build a project management tool at the web application company 37signals, extracted the framework he had created and released it as open source. He called it Ruby on Rails, and Rails demonstrated what Matsumoto had only promised: that Ruby could build full-scale web applications with astonishing speed and elegance. The qualities Matsumoto had designed in from the start, flexibility, expressiveness, a willingness to let you write code that reads almost like English, turned out to be exactly what the web needed.

The world paid attention then. By 2006, Ruby on Rails had become a phenomenon. Startup companies adopted Rails with enthusiasm. The language that Matsumoto had designed in 1993 with no commercial backing, driven purely by a vision of programmer happiness, had suddenly become the language of the web boom. But that vision never changed. Ruby remained what it was built to be: a language that believed the programmer mattered, that assumed you were smart and trusted you to write what you meant, and that had been shaped, from its very name onward, to wish you well.

What happened next

Year	Event

1993	Matsumoto begins developing Ruby; by August, achieves first "Hello world" program
1995	Ruby 0.95 released publicly on December 21; initial mailing list launched
1997	Matsumoto hired by Netlab to work on Ruby full-time
2000	Ruby surpasses Python in usage within Japan; Dave Thomas publishes "Programming Ruby," the first major English-language book
2004	David Heinemeier Hansson releases Ruby on Rails as open source in July; community motto MINASWAN established
2005	Rails open source with full commit rights; Heinemeier Hansson named "Hacker of the Year" by Google and O'Reilly
2006	Ruby on Rails becomes a mainstream phenomenon; Ruby adoption accelerates globally
2010	Ruby reaches 1.9.2, becoming significantly faster and more mature

A SPOTLIGHT ON MALBOLGE

The Impossible Program

Ben Olmstead Built a Language No One Could Use. Someone Used It Anyway.

In 1998, a computer science student named Ben Olmstead set out to create the most difficult programming language ever designed. He named it Malbolge, after the eighth circle of Hell in Dante's Inferno, the circle reserved for fraudsters and deceivers.

Olmstead didn't just want Malbolge to be hard. He wanted it to be impossible. Malbolge runs on a base-3 virtual machine with three registers. A single instruction does different things depending on its memory address, and every instruction rewrites itself after running, so the program you write is not the program that executes. The difficulty isn't a side effect of the design. It is the entire point.

Olmstead was right about the difficulty. He himself could not write a Hello World program in his own language. Neither could anyone else, for nearly two years.

Then, in 2000, Andrew Cooke cracked it. He didn't write the program by hand. He wrote a beam search algorithm, a form of artificial intelligence used in speech recognition and machine translation, that explored billions of possible Malbolge programs and found one that produced the correct output. A computer had to write the program because no human could.

The resulting Hello World is a string of seemingly random symbols:

```
(=<`#9]~6ZY327Uv4-QsSNq.K+Jh)d
%e`#th4[Si:Krjow_G'"E#}Cb@!>x*9?2V
,Dp0Ij<NO&fR,+<v%$@M6$1
```

Malbolge Hello World, found by beam search, not written by a human

There is no structure a human can read. There is no logic you can follow. Every character encodes an operation that depends on its position in memory and mutates after execution. The program is, by design, indistinguishable from noise.

Malbolge's existence raises a question that sounds like philosophy but is actually computer science: what is the minimum threshold for a usable programming language? The original Malbolge isn't even Turing-complete, its fixed memory size limits what it can compute. A later variant, Malbolge Unshackled, removed that constraint in 2007. But either way, Brainfuck's eight instructions prove that Turing completeness requires almost nothing. Malbolge proves that having everything, instructions, memory, registers, doesn't mean you can actually use it.

In the years since, a small community has formed around Malbolge. Researchers have learned to write programs (carefully, painfully) by reasoning about the self-modifying instruction cycles. Lou Scheffer published an analysis showing that certain instruction sequences stabilize into predictable patterns, creating "islands of sanity" in the chaos. Others have written Malbolge programs that perform arithmetic, sort lists, and even play simple games.

Olmstead later created Dis, another language named after Dante's circles, somewhat less hostile than its predecessor. He never expected anyone to actually use Malbolge. But programmers, it turns out, are exactly the kind of people who see an impossible challenge and refuse to leave it alone. Malbolge was designed as a locked room with no key. The community picked the lock anyway.

What happened next

Year	Event
1972	INTERCAL created as the first deliberately perverse programming language
1993	Urban Muller creates Brainfuck with eight commands and a compiler under 300 bytes
1998	Ben Olmstead designs Malbolge to be nearly impossible to program in
2000	Andrew Cooke generates the first Malbolge program using beam search in Lisp
2001	David Morgan-Mar creates Piet, where programs are abstract paintings
2003	Whitespace released; only spaces, tabs, and line feeds carry meaning
2007	GolfScript released, launching competitive code golf as a subculture
2012	The Esolang Wiki becomes the central repository for esoteric languages
2015	Jelly, 05AB1E, and other golfing languages emerge; esolang community grows

Era 6

The Modern Era

2000–2009

For forty years, programming languages competed on how fast they could make a machine run. In the 2000s, they began competing on how fast they could make a programmer think. Hardware had finally gotten fast enough that the bottleneck shifted from the machine to the human, and a new generation of languages optimized for developer productivity emerged.

C# was Microsoft's answer to Java, designed by Anders Hejlsberg, the same mind behind Turbo Pascal and Delphi. Scala attempted what Java couldn't: a true fusion of object-oriented and functional programming. F# proved that functional programming could thrive on Microsoft's .NET platform. Meanwhile, CUDA revolutionized parallel computing with little fanfare, letting programmers harness thousands of GPU cores through simple C extensions, laying the hardware foundation for the AI revolution a decade later.

The iPhone launched in 2007, creating entirely new categories of software. Scratch, released the same year, proved that anyone, even children, could program. CoffeeScript challenged JavaScript's verbosity. Groovy brought scripting to the JVM. Nim and Go tackled systems programming from opposite directions. By the end of the decade, the diversity of successful languages was remarkable: each specialized for different domains and platforms, unified by a shared commitment to making programmers more productive.

11 languages in this era

Timeline

- 2000 — Anders Hejlsberg and Microsoft release C# and .NET
- 2003 — James Strachan and Bob McWhirter create Groovy
- 2004 — Martin Odersky creates Scala at EPFL
- 2005 — Don Syme creates F# at Microsoft Research
- 2006 — Jeffrey Snover releases PowerShell
- 2007 — Mitchel Resnick launches Scratch at MIT Media Lab; Rich Hickey creates Clojure; NVIDIA releases CUDA SDK
- 2008 — Andreas Rumpf creates Nim
- 2009 — Rob Pike, Ken Thompson, and Robert Griesemer create Go at Google

World Context

Dot-com crash aftermath; rise of Web 2.0; smartphones beginning with iPhone (2007); cloud computing emerging

Dominant Paradigms

Dynamic languages mainstream; functional programming resurgence on JVM; GPU computing with CUDA; languages for specific domains thrive

A SPOTLIGHT ON C#

The Language Born from a Lawsuit

Anders Hejlsberg, a Court Order in San Jose, and the Making of C#

In late 1996, Microsoft paid Anders Hejlsberg roughly a million dollars to leave Borland. He was not the only one. Over thirty months, Microsoft hired thirty-four of Borland's top engineers with signing bonuses that dwarfed their annual salaries. Borland sued, calling it a coordinated raid designed to cripple a competitor. The case settled in four months. By then the damage was done; Borland had lost the architect of Turbo Pascal and Delphi, two of the most beloved programming tools ever built.

Hejlsberg had written Turbo Pascal at twenty-two, working alone in Copenhagen. It sold for $49.95 when competing compilers cost hundreds, and Borland expected thirty thousand copies. They sold over four hundred thousand. The secret was speed: Hejlsberg's compiler worked entirely in memory, producing executables without touching the disk. It pioneered the integrated development environment, edit, compile, debug, and run in one place. As he later put it: "When you typed your code, you wanted to run it immediately." That instinct for fast feedback would shape every language he touched.

Microsoft put him to work on Visual J++, their proprietary Java implementation. The problem was that Microsoft's version of Java wasn't really Java. It included extensions that let programs call Windows system libraries directly, bypassing the cross-platform promise that was Java's entire reason for existing. Sun Microsystems had licensed Java to Microsoft with a compatibility requirement. Microsoft shipped it with the "Java Compatible" logo anyway, despite failing Sun's own test suites.

Sun sued in October 1997. A federal judge in San Jose barred Microsoft from using the Java trademark, then issued a broader order requiring compliance with Sun's compatibility standards. By January 2001, Microsoft paid twenty million dollars, had its Java license terminated, and was frozen at Java 1.1.4, prohibited from further Java development. The product Hejlsberg had spent years building was dead.

But the court order didn't kill the idea. It freed it. As Hejlsberg explained years later: "The product that we built, Visual J++ 6.0, ended up earning Microsoft a court order from a judge in San Jose. So in a sense, all of that was the genesis for what came next, which was .NET and C#, because Microsoft realized that building a future platform on top of something licensed from another partner wasn't going to work."

In December 1998, while the lawsuit was still raging, Hejlsberg started designing a new language under the code name COOL, C-like Object Oriented Language. He assembled a small team that met every Monday, Wednesday, and Friday afternoon. Microsoft considered keeping the name but couldn't trademark it. They tried e-C, Safe C, C-square, and Cesium before settling on C#, a musical sharp sign raising C by a semitone, and a visual pun: four plus signs in a grid, C++ incremented.

C# launched in February 2002 and was immediately dismissed as a Java clone. The structural similarities were obvious: garbage collection, single inheritance, interfaces, familiar syntax. But Hejlsberg had studied Java's limitations and designed around them. By version 2.0, C# had generics with full runtime support where Java used type erasure. By 3.0, LINQ let developers write database queries in language syntax, a feature inspired by functional programming that Java wouldn't match for years. By 5.0, async/await made asynchronous code readable, years ahead of the competition.

Then came the reversal nobody at Microsoft had anticipated. Unity, the game engine, had supported three scripting languages. It dropped them one by one until C# was the sole survivor, and no corporate mandate drove the choice; developers simply preferred it. At its peak, by Unity's own count, half of all new games were built on it. The language Microsoft designed for enterprise Windows development became how an entire generation learned to code, by building games.

The final act arrived under Satya Nadella, who became CEO in 2014 and began dismantling the combative culture that had produced the Java lawsuit in the first place. Microsoft open-sourced the entire C# compiler, moved it to GitHub, and released .NET Core, a fully cross-platform runtime for Windows, Linux, and macOS. The company that had once called Linux a "cancer" was now running its flagship language on it.

Meanwhile, Hejlsberg had already started his next project. In 2012, he began building TypeScript, a superset of JavaScript that added static types to the language that runs the web. By 2025, TypeScript was the most-used language on GitHub, overtaking both JavaScript and Python. Four decades, four languages (Turbo Pascal, Delphi, C#, TypeScript), each one a response to the same instinct: make the feedback loop faster, make the programmer more productive, and ship it before anyone says it can't be done.

What happened next

2002	C# 1.0 launches with Visual Studio .NET. Critics call it a Java clone.
2005	C# 2.0 introduces reified generics, technically superior to Java's approach.
2007	C# 3.0 adds LINQ and lambdas. The "Java clone" label stops sticking.
2012	Hejlsberg begins TypeScript. C# 5.0 introduces async/await.
2014	Nadella becomes CEO. Microsoft open-sources the Roslyn compiler.
2016	.NET Core 1.0: fully open-source and cross-platform.
2017	Unity drops all scripting languages except C#.
2025	TypeScript becomes the most-used language on GitHub.

C# 2000

Anders Hejlsberg, Microsoft ● Active

Anders Hejlsberg had already created Turbo Pascal and Delphi at Borland , two of the most beloved programming tools ever made. When he left for Microsoft, Borland sued. Microsoft put him in charge of building Java's competitor, and C# became the cornerstone of the .NET ecosystem. C# evolved into one of the most feature-rich modern languages, with LINQ, async/await, and pattern matching.

```
using System;
class HelloWorld {
  static void Main() {
    Console.WriteLine("Hello, World!");
  }
}
```

using directives; Console.WriteLine outputs

Paradigm: Object-Oriented

Typing: Static

Usage: ●●●●○

Lines: 6

★ C# with LINQ enabled developers to write database queries in language syntax; it shaped how Python and JavaScript handle data.

GROOVY 2003

James Strachan, Bob McWhirter ● Active

James Strachan wrote a blog post in 2003 complaining that Java was too verbose for scripting tasks. Then he decided to build the fix himself. Groovy ran on the JVM, gave developers Python-like ease with full access to Java's ecosystem, and quietly powered Android build systems (via Gradle) for a decade. It proved that the JVM could host languages far more expressive than Java itself.

```
println 'Hello, World!'
```

Python-like syntax; println outputs

Paradigm: Multi-paradigm

Typing: Dynamic

Usage: ●●○○○

Lines: 1

★ Gradle originally used Groovy for its build scripts, and every Android app on Earth was configured in Groovy DSL for years, until Kotlin DSL became the default in 2023, ending Groovy's most visible role.

SCALA 2004

Martin Odersky, École Polytechnique Fédérale de Lausanne ● Active

Martin Odersky had already helped design Java generics, and the experience convinced him that Java's fundamental design made it impossible to properly unify object-oriented and functional programming. So he built a language that could. Scala runs on the JVM with full Java interoperability, but adds pattern matching, immutability, and type inference. Twitter, LinkedIn, and Netflix bet their backends on it.

```
object HelloWorld extends App {
  println("Hello, World!")
}
```

Object declaration; extends App for executable

Paradigm: Multi-paradigm

Typing: Static

Usage: ●●○○○

Lines: 3

★ Twitter rebuilt its backend from Ruby to Scala to handle the 2010 World Cup traffic spikes, a migration so successful that Scala became synonymous with high-scale JVM engineering. LinkedIn and Netflix followed.

F# 2005

Don Syme, Microsoft Research ● Active

Don Syme spent years at Microsoft Research proving that functional programming could work on .NET. Microsoft management was skeptical , functional languages were supposed to be academic curiosities, not production tools. Syme proved them wrong, and F# became the language of choice for financial engineers writing trading algorithms. F# pioneered type providers and computation expressions that later influenced C#'s own evolution.

```
printfn "Hello, World!"
```

printfn with formatted output

Paradigm: Functional

Typing: Static with Inference

Usage: ●●○○○

Lines: 1

★ F# is so beloved by financial engineers that major banks use it for trading algorithms; it combines safety with performance.

A SPOTLIGHT ON F#

The Waiting Room

Don Syme and the Feature That Spread to Nearly Every Language

In April 2007, Don Syme traveled from Microsoft Research in Cambridge to EPFL in Lausanne for a six-week sabbatical with Martin Odersky. Odersky was building Scala. Syme was building F#. Both men were trying to make functional programming practical on industry platforms, and they had been exchanging ideas for years. During the sabbatical, Syme saw a construct that one of Odersky's students, Philipp Haller, had built: `react { ... }`, a primitive for message processing. It crystallized a problem Syme had been thinking about. Asynchronous programming needed something built into the language, not bolted on as a library.

Back in Cambridge, Syme added `async { ... }` to F#. A programmer could take any block of synchronous code, wrap it in `async { }`, and replace `let` with `let!` at the points that needed to wait. The compiler handled the rest, rewriting clean, readable source into the state-machine logic that asynchronous execution actually requires. On October 10, 2007, Syme announced the feature in a blog post titled "Introducing F# Asynchronous Workflows."

People noticed. At Microsoft's Professional Developers Conference in 2008, Luca Bolognese gave a packed keynote on async that had the audience gasping. The F# team gave hundreds of talks as they brought the language to market in Visual Studio 2010. Every one featured async. Within the Microsoft developer world, the pressure to "do something about parallel and async" built and built. But thousands of people noticing is still small. You need one or two extra zeros. C# had those zeros. Its curly-brace syntax was what most of the industry was already familiar with. Syme and his colleagues presented the pattern to the C# design team. It took five years.

In 2012, C# 5.0 shipped with `async` and `await`. The feature became one of the most celebrated additions in the language's history. Millions of developers adopted it. Then Python added `async`/`await` in version 3.5. Then JavaScript, in ES2017. Then Kotlin. Then Rust. Then Swift. By 2020, having an async modality in a language was effectively an industry standard, and in each case the lineage traced back, at least partly, through F#.

Async was not the beginning of the story. Syme had started F# around 2002, drawing on OCaml as its foundation. Before F# existed, he had already designed and implemented generics for the .NET Common Language Runtime, the feature that C# and Visual Basic would depend on for the next two decades. He understood Microsoft's platform from the inside. But F# itself was a harder sell. Microsoft management saw functional languages as academic curiosities, not production tools.

The language grew anyway. Financial engineers at major banks adopted it for trading systems. Scientists used it. A small, devoted community formed around the idea that a typed functional language could be practical enough to use at work. F# pioneered type providers, which let the compiler connect directly to external data sources and generate types automatically. It introduced computation expressions, a general mechanism for extending the language's syntax. Each feature proved the same point: functional programming ideas, delivered in a practical package, could improve any language.

The people around F# proved it too. Simon Peyton Jones, one of the architects of Haskell, worked at Microsoft Research and advised Syme directly on active patterns and extensible pattern matching. The exchange with Odersky in Lausanne ran both ways: F#'s active patterns influenced Scala's extractors, just as Scala's work had helped inspire F#'s async design.

Then the people started moving to other languages. Luke Hoban worked on F# 2.0, then became one of the co-originators of TypeScript. Syme later wrote that TypeScript's type checking and inference "would not have appeared from a Microsoft team in anything close to its current form without the influence of F#." Joe Pamer moved from the F# community to Apple, where he managed the Swift compiler team from 2014 to 2016. Wes Dyer, influenced by F#'s first-class events, helped create the Reactive Extensions project, which spread across multiple languages.

The influence on C# ran especially deep. F#'s type inference shaped C# 3.0. Its async workflows became C# 5.0's async/await. Its tuples and pattern matching appeared in C# 7.0. Its non-nullness by default, a design choice Syme made from the outset, preceded C# 9.0's adoption of the same principle by over a decade. Given that C# is one of the most widely used languages in the world, F#'s fingerprints are everywhere, even where its name is not.

In 2010, F# went open source, years before Microsoft's broader embrace of open-source development. Syme had understood before most of his colleagues that a language lives or dies by its community, not its corporate sponsor. F# was used to build a quantum computing simulator, to model the cladding of the Louvre Abu Dhabi, and to power Jet.com, the startup that Walmart acquired for over three billion dollars.

The feature Syme introduced in that 2007 blog post now runs inside nearly every major programming language on earth. He spent more than twenty years building a language that the industry kept borrowing from without always knowing the source. He is still building it.

What happened next

Year	Event
2001	Don Syme begins work on .NET generics at Microsoft Research Cambridge; the implementation ships with .NET 2.0 and becomes foundational to C# and Visual Basic
2002	Syme starts building F# on top of OCaml's core, targeting the .NET Common Language Runtime
2005	F# 1.0 released at Microsoft Research; introduces a functional-first language on .NET with type inference, pattern matching, and non-null types by default
2007	Syme introduces async workflows during a sabbatical at EPFL; announces "Introducing F# Asynchronous Workflows" on October 10
2010	Luke Hoban, after working on F# 2.0, co-originates TypeScript (initially called Strada) at Microsoft
2010	F# goes open source under the Apache 2.0 license, years before Microsoft's broader open-source shift; F# 2.0 ships with Visual Studio 2010
2012	F# 3.0 introduces type providers; C# 5.0 ships with async/await, directly influenced by F#'s async workflows
2014	Joe Pamer joins Apple to manage the Swift compiler team (2014-2016)
2015	Python 3.5 adds async/await; F# 4.0 released
2016	Jet.com, built largely on F#, acquired by Walmart for over three billion dollars
2017	JavaScript adds async/await in ES2017; Kotlin adds coroutines
2019	Rust stabilizes async/await in version 1.39
2020	Syme publishes "The Early History of F#" as a HOPL paper; async/await is now an industry standard across major languages

POWERSHELL 2006

Jeffrey Snover, Microsoft ● Active

Jeffrey Snover created PowerShell after writing a manifesto arguing that Windows desperately needed a real command-line shell. Unlike Unix shells that pipe text, PowerShell pipes .NET objects, meaning you can manipulate structured data without parsing strings. Microsoft initially resisted the idea, but PowerShell became essential for Windows system administration and eventually went cross-platform on Linux and macOS.

```
Write-Host "Hello, World!"
```

Write-Host outputs to console; PowerShell uses verb-noun command naming

Paradigm: Multi-paradigm

Typing: Dynamic

Usage: ●●●●○

Lines: 1

★ Snover's original manifesto was rejected by Microsoft management. He kept pushing for years until they let him build it; PowerShell shipped with every Windows installation from Vista onward and managed millions of servers worldwide.

CUDA 2007

Ian Buck, NVIDIA ● Active

Ian Buck led the development of CUDA at NVIDIA to let programmers harness GPU parallel processing for general computation. Before CUDA, GPUs could only render graphics. CUDA opened them to scientific simulation, cryptography, and eventually deep learning, by 2025 every major neural network trained on CUDA. It turned NVIDIA from a graphics card company into the most valuable company on Earth.

```
#include <cstdio>
__global__ void hello() {
  printf("Hello, World!\n");
}
int main() {
  hello<<<1,1>>>();
}
```

__global__ marks a GPU kernel; <<<1,1>>> launches one thread on one block

Paradigm: Imperative

Typing: Static

Usage: ●●●●○

Lines: 7

★ CUDA's triple-angle-bracket syntax (<<<blocks, threads>>>) may be the most valuable punctuation in computing history; it helped unlock a multi-trillion-dollar AI hardware market.

CLOJURE 2007

Rich Hickey ● Active

Rich Hickey created Clojure as a modern Lisp for the JVM built around immutability and functional programming. Clojure runs on the JVM but brings Lisp's power to Java developers. Its focus on immutable data structures changed how developers approach concurrent programming.

```
(println "Hello, World!")
```

Lisp-style; println outputs

Paradigm: Functional

Typing: Dynamic

Usage: ●●○○○

Lines: 1

★ Clojure's immutable-first approach influenced a generation of developers, demonstrating that embracing immutability prevents entire classes of bugs.

SCRATCH 2007

Mitchel Resnick, MIT Media Lab ● Active

Mitchel Resnick at MIT Media Lab created Scratch as a visual programming language for learning. Instead of typing code, students drag and drop blocks representing programming concepts. Scratch's intuitive interface has taught millions of children to code; it democratized programming education.

```
say [Hello, World!]
```

Block-based; say block outputs to sprite

Paradigm: Imperative

Typing: Dynamic

Usage: ●●●○○

Lines: 1

★ Scratch crossed one billion total projects in April 2024, with over 160 million shared publicly. The platform had more registered users than the population of Japan.

NIM 2008

Andreas Rumpf ● Active

Andreas Rumpf started Nim alone in 2008, frustrated that no language gave him Python's readability with C's performance. He spent over a decade as a solo developer, refining a compiler that outputs C, C++, or JavaScript from the same source, before the community finally caught up to his vision. Nim compiles to efficient native code while offering a clean, Python-like syntax.

```
echo "Hello, World!"
```

echo compiles to efficient C; Nim's syntax is Python-like but statically typed

Paradigm: Multi-paradigm

Typing: Static with Inference

Usage: ●●○○○

Lines: 1

★ Nim compiles to C, C++, or JavaScript, meaning a single Nim program can target everything from embedded hardware to web browsers.

COFFEESCRIPT 2009

Jeremy Ashkenas ◐ Endangered

Jeremy Ashkenas created CoffeeScript as a language that compiles to JavaScript, bringing Python-like syntax to the browser. CoffeeScript showed what transpilation could do for JavaScript. Its ideas influenced modern JavaScript (ES6+ arrow functions, destructuring).

```
console.log "Hello, World!"
```

Python-like syntax; compiles to JavaScript

Paradigm: Multi-paradigm

Typing: Dynamic

Usage: ●●○○○

Lines: 1

★ Dropbox's entire web frontend was written in CoffeeScript at its peak. When ES6 adopted arrow functions and destructuring, ideas CoffeeScript pioneered, CoffeeScript essentially won by making itself obsolete.

GO 2009

Rob Pike, Ken Thompson, Robert Griesemer, Google ● Active

Rob Pike, Ken Thompson, and Robert Griesemer at Google created Go as a simple, fast language for systems programming in the cloud era. Go's concurrency model with goroutines made it easy to write concurrent programs. It became the language of infrastructure; Docker, Kubernetes, and cloud tools were built in Go.

```
package main
import "fmt"
func main() {
  fmt.Println("Hello, World!")
}
```

Package-based; fmt for formatted output

Paradigm: Imperative

Typing: Static

Usage: ●●●●○

Lines: 5

★ Go's simplicity and fast compilation made it the language of cloud infrastructure, powering Docker and containerization systems.

A SPOTLIGHT ON CUDA

The Graphics Card That Ate the World

CUDA, NVIDIA, and the Hardware Foundation of the AI Revolution

In 2003, a Stanford PhD student named Ian Buck walked into the office of David Kirk, NVIDIA's chief scientist, with a question that sounded slightly insane: what if we stopped using graphics cards just for graphics?

The idea wasn't entirely new. Researchers had been exploiting GPU hardware for non-graphical computation since the early 2000s, a practice called GPGPU, general-purpose computing on graphics processing units. But the process was absurd: to multiply two matrices, you had to disguise the numbers as pixel colors, feed them through a rendering pipeline designed for video games, and extract the results from what the GPU thought was an image. It worked, barely. It was also completely impractical for real-world use.

Buck's insight, refined over years of academic work and his Brook language prototype, was that GPUs didn't need to be tricked into doing general computation. They needed to be designed for it. A GPU has thousands of simple cores running in parallel, vastly more than any CPU. If you could give programmers direct access to those cores with a language they already knew, the performance gains would be transformative.

NVIDIA hired Buck and gave him a team. In early 2007, they released CUDA (Compute Unified Device Architecture), a set of extensions to C that let programmers write code for the GPU almost as easily as they wrote code for the CPU. The key syntax was the triple-angle-bracket kernel launch:

```
__global__ void hello() {
    printf("Hello, World!\n");
}

int main() {
    hello<<<1, 1>>>();
}
```

The triple-angle-bracket <<<>>> kernel launch, CUDA's signature syntax

Those angle brackets told the GPU to run a function across thousands of threads simultaneously. It was, in hindsight, the most commercially valuable punctuation mark ever invented.

The first adopters were scientists. Molecular dynamics simulations that took weeks on CPUs ran in hours on GPUs. Weather models, protein folding, financial Monte Carlo simulations, anywhere the same computation needed to happen millions of times, CUDA delivered ten- to hundred-fold speedups. But the real earthquake came from an unexpected direction.

In 2012, a team at the University of Toronto used CUDA-accelerated GPUs to train a deep neural network called AlexNet. It won the ImageNet competition by a margin so large it effectively ended the previous era of computer vision overnight. The deep learning revolution had begun, and it ran on NVIDIA hardware.

Everything that followed, GPT, DALL-E, self-driving cars, protein structure prediction, depended on the ability to train neural networks on thousands of GPU cores in parallel. CUDA was the software layer that made it possible. NVIDIA's market capitalization went from roughly $7 billion in 2012 to over $4.5 trillion by early 2026, becoming the most valuable company on Earth. A graphics card company had become the backbone of artificial intelligence.

Ian Buck became NVIDIA's VP of accelerated computing. The triple-angle-bracket syntax he helped design became a fixture of every serious machine learning course. And the question he asked David Kirk in 2003 (what if graphics cards did more than graphics?) became the foundation of the most consequential hardware revolution since the microprocessor.

What happened next

Year	Event
2003	Ian Buck develops Brook at Stanford, showing GPUs can run general-purpose computations
2007	NVIDIA releases CUDA 1.0, giving developers direct access to GPU parallel processing
2010	Fermi architecture ships with C++ support and 512 stream processors
2012	AlexNet wins ImageNet trained on two GTX 580 GPUs; deep learning era begins
2016	NVIDIA ships the P100, the first GPU purpose-built for deep learning
2023	NVIDIA surpasses $1 trillion market cap, driven by AI training demand
2024	NVIDIA reaches $2 trillion; H100 dominates AI training; Blackwell announced

Era 7

The Safety Era

2010–2019

The 2010s were defined by a singular obsession: safety. Languages competed fiercely on how many bugs they could prevent without sacrificing expressiveness. Rust demonstrated that memory safety and performance could coexist without a garbage collector. TypeScript proved that type systems could enhance JavaScript without fundamentally changing it. Swift replaced Objective-C for Apple development with a language designed to eliminate entire classes of common errors.

Mobile platforms drove language innovation too, with Kotlin becoming the preferred language for Android and Swift for iOS. The decade showed that language adoption isn't about technical purity, it's about platforms, communities, and ecosystems. Julia tackled the 'two-language problem' in scientific computing. Elixir brought functional programming to web development with Erlang's battle-tested reliability underneath. Zig offered a modern alternative to C without hidden control flow. And Raku, the language formerly known as Perl 6, finally shipped after fifteen years of development. By 2019, the diversity of successful languages was remarkable: each specialized for different domains and platforms, unified by a shared commitment to preventing bugs and improving developer experience.

13 languages in this era

Timeline

- 2010 — Graydon Hoare releases Rust at Mozilla
- 2011 — JetBrains creates Kotlin; Google releases Dart; José Valim creates Elixir
- 2012 — Anders Hejlsberg creates TypeScript; Julia releases v0.1
- 2014 — Chris Lattner releases Swift for Apple platforms
- 2016 — Andrew Kelley creates Zig
- 2019 — Perl 6 officially renamed Raku

World Context

Rise of machine learning; containerization revolution; smartphone ubiquity; growing concern about data privacy

Dominant Paradigms

Memory safety paramount; type systems sophisticated; multiple paradigms coexist; mobile-first development; specialized languages for specific domains

RUST 2010

Graydon Hoare, Mozilla ● Active

Graydon Hoare at Mozilla created Rust as a systems language combining memory safety with performance. Rust's ownership system prevents entire categories of bugs (memory leaks, data races, null pointer dereferences) without garbage collection. It's become the safest systems language, used by Linux, AWS, and Chrome.

```
fn main() {
    println!("Hello, World!");
}
```

fn defines functions; println! is macro

Paradigm: Multi-paradigm

Typing: Static

Usage: ●●●●○

Lines: 3

★ In 2024, the White House Office of the National Cyber Director recommended Rust by name as a memory-safe alternative to C and C++, the first time a US government agency endorsed a specific programming language for cybersecurity.

DART 2011

Lars Bak, Kasper Lund, Google ● Active

Google originally designed Dart to replace JavaScript in Chrome, they even built a Dart VM into the browser. When that plan was abandoned, Dart seemed dead. Then Flutter arrived, and Dart found a larger audience in mobile development than the original browser vision ever would have delivered. Dart compiled to JavaScript for the browser and to native code for servers, becoming the foundation for mobile app development.

```
void main() {
  print('Hello, World!');
}
```

void main entry point; print outputs

Paradigm: Object-Oriented

Typing: Static

Usage: ●●●○○

Lines: 3

★ Dart was originally designed to replace JavaScript in browsers, Google even built a Dart VM into Chrome. When that plan was abandoned, Flutter gave Dart a second life as a mobile framework, and it found a larger audience than the original vision ever would have.

ELIXIR 2011

José Valim ● Active

José Valim created Elixir as a functional language for the Erlang Virtual Machine, combining Erlang's fault tolerance with a modern syntax. Elixir's pipe operator and macros made concurrent programming accessible. It's used by Discord, Pinterest, and companies building resilient systems.

```
IO.puts "Hello, World!"
```

IO.puts outputs; pipe operator enables composition

Paradigm: Functional

Typing: Dynamic

Usage: ●●○○○

Lines: 1

★ By 2019, Discord was using Elixir to handle over 11 million concurrent users across its platform. Elixir's concurrency model, inherited from Erlang, scaled where Node.js and Python couldn't.

IDRIS 2011

Edwin Brady, University of St Andrews ● Active

What if your type system was so powerful it could mathematically prove your code was correct, before you ran it? Edwin Brady built Idris to answer that question. In Idris, a function's type signature can guarantee properties like "this list-append always returns a list whose length equals the sum of the inputs." If the math doesn't check out, the code won't compile. Types become proofs, and the compiler becomes a theorem checker.

```
main : IO ()
main = putStrLn "Hello, World!"
```

Type signature; IO () is I/O action

Paradigm: Functional

Typing: Static with Dependent Types

Usage: ●○○○○

Lines: 2

★ In Idris, you can write a type signature that guarantees a list-append function returns a list whose length equals the sum of the two inputs. The program won't compile if the math doesn't check out.

KOTLIN 2011

JetBrains ● Active

JetBrains created Kotlin as a modern, pragmatic language for the JVM, learning from Java's struggles. Kotlin eliminated null pointer exceptions through nullable types and added extension functions, coroutines, and data classes. Google officially endorsed Kotlin as the preferred language for Android development.

```
fun main() {
    println("Hello, World!")
}
```

fun defines functions; println outputs

Paradigm: Multi-paradigm

Typing: Static

Usage: ●●●○○

Lines: 3

★ Google announced Kotlin as a first-class Android language in 2017, and by 2019 declared it the preferred language, displacing Java after a decade of Android dominance. JetBrains, Kotlin's creator, isn't a platform company; it makes IDEs.

ELM 2012

Evan Czaplicki ● Active

Evan Czaplicki created Elm while working on his senior thesis. His goal sounded impossible, a front-end language that produces zero runtime exceptions. Elm compiles to JavaScript and prevents null pointer exceptions, type errors, and other runtime errors through its type system. Companies deployed hundreds of thousands of lines of Elm and reported exactly that: no crashes. It inspired the architecture of frameworks like Redux.

```
import Html exposing (Html, text)
main : Html msg
main = text "Hello, World!"
```

Type signature; text renders text

Paradigm: Functional

Typing: Static

Usage: ●●○○○

Lines: 3

★ Elm's compiler famously produces zero runtime exceptions in production. The Elm community tracked this for years, companies reported hundreds of thousands of lines of Elm code running without a single runtime crash.

JULIA 2012

Jeff Bezanson, Stefan Karpinski, Viral Shah, Alan Edelman, MIT ● Active

MIT researchers created Julia as a language for numerical and scientific computing, combining the ease of Python with the speed of C. Julia's multiple dispatch and just-in-time compilation made it formidable for mathematical programming. It's becoming the standard for machine learning and scientific computing.

```
println("Hello, World!")
```

println outputs with newline

Paradigm: Multi-paradigm

Typing: Dynamic

Usage: ●●○○○

Lines: 1

★ Julia's JIT compiler let Celeste.jl simulate the entire Earth's climate system at unprecedented resolution. Scientists could write code as readable as Python that ran within a factor of two of hand-optimized C.

TYPESCRIPT 2012

Anders Hejlsberg, Microsoft ● Active

Anders Hejlsberg at Microsoft created TypeScript as a typed superset of JavaScript. TypeScript adds static types while remaining fully compatible with JavaScript, creating a sweet spot between dynamic and static typing. It became the standard for large-scale JavaScript applications.

```
console.log('Hello, World!');
```

Compiles to JavaScript

Paradigm: Multi-paradigm

Typing: Static

Usage: ●●●●○

Lines: 1

★ By 2026, TypeScript appeared in the majority of new JavaScript projects on GitHub. Google, Microsoft, Airbnb, and Stripe all adopted it for large codebases; the type system caught classes of bugs that JavaScript's dynamic nature hid until production.

CRYSTAL 2014

Ary Borenszweig, Juan Wajnerman, others ● Active

A group of Argentine developers loved Ruby's syntax but needed compiled performance for production systems. They set out to prove you could have both, and built Crystal, a language that compiles Ruby-like code to fast native binaries with full static type inference. It's becoming popular for systems programming and microservices.

```
puts "Hello, World!"
```

Ruby-like syntax; compiles to machine code

Paradigm: Multi-paradigm

Typing: Static with Inference

Usage: ●●○○○

Lines: 1

★ Crystal compiles Ruby's syntax to fast native code, combining the joy of Ruby with the performance of C.

HACK 2014

Facebook ● Active

Facebook created Hack as a statically typed dialect of PHP with gradual typing. Hack lets developers add types to existing PHP code gradually, improving safety without requiring a complete rewrite. It powers Facebook's infrastructure.

```
<?hh
<<__EntryPoint>>
function main(): void {
  echo "Hello, World!";
}
```

<<__EntryPoint>> marks entry function; echo outputs

Paradigm: Imperative

Typing: Static (Gradual)

Usage: ●○○○○

Lines: 5

★ Facebook migrated its entire PHP codebase (millions of lines) to Hack incrementally, file by file. Engineers could add types to one function while the untyped code next to it kept running. No big-bang rewrite required.

SWIFT 2014

Chris Lattner, Apple ● Active

Chris Lattner at Apple created Swift as a modern replacement for Objective-C. Swift combined Python's readability with strong typing, safety features, and performance. Apple adopted it as the primary language for iOS, macOS, watchOS, and tvOS development.

```
print("Hello, World!")
```

No main function needed, Swift infers the entry point from top-level code

Paradigm: Multi-paradigm

Typing: Static

Usage: ●●●○○

Lines: 1

★ Swift's adoption by Apple meant millions of developers switched languages overnight. Platform support, it turned out, drives adoption more than features.

ZIG 2016

Andrew Kelley ● Active

Andrew Kelley created Zig as a minimal systems language built for clarity, safety, and performance. Zig has no hidden control flow, making it predictable for systems programming. It's gaining adoption for competing with C and Rust in performance-critical applications.

```
const std = @import("std");
pub fn main() !void {
  const stdout = std.io.getStdOut().writer();
  try stdout.print("Hello, World!\n", .{});
}
```

@import for standard library; stdout writer for output

Paradigm: Imperative

Typing: Static

Usage: ●●○○○

Lines: 5

★ Zig has no hidden control flow and no hidden allocations. Andrew Kelley demonstrated that C's 'simple' for loop hides branch prediction, stack manipulation, and implicit conversions, Zig makes every cost visible in the source code.

A SPOTLIGHT ON ZIG

The Donation Button

Andrew Kelley, a Mass Resignation from Open Source, and the True Cost of Free Software

In June 2018, Andrew Kelley did something most people only dream about: he walked away from a six-figure salary to bet his life on strangers' goodwill. Not on venture capital. Not on a startup with a business model. On donations. On the kindness of a community that used a programming language he had started on nights and weekends, in stolen hours at his previous employer.

Kelley was a senior backend software engineer at OkCupid, maintaining a massive C++ codebase. He had negotiated fiercely during his hiring to retain ownership of code he wrote in his spare time, this would become Zig. For years, he had balanced his day job with his evening passion project, watching the language grow from a private obsession into a real community. But the growth became a problem. Some weekends, he wrote later, "I found myself with only enough time to merge pull requests and respond to issues, and no time leftover to make progress on big roadmap changes." Forty hours of legacy code by day, and then a mailbox overflowing with the language's emergent users and their requests, their bugs, their expectations. The equation no longer balanced.

The crisis had nothing to do with technical vision. It was a crisis of human attention. Zig had become too big for its creator to sustain as a hobby, but not yet large enough to fund itself. Kelley faced a choice: abandon Zig to the natural entropy of unmaintained open source, or sacrifice the financial security that most engineers take as a given. He chose the third path, the one almost nobody takes. He bet his mortgage, his health insurance, his retirement account on the belief that people who loved what Zig represented would pay for its continued existence.

Kelley's decision arrived as an aftershock to a much deeper tremor running through software development. For years, critical infrastructure had been maintained by unpaid volunteers while billion-dollar companies built their entire operations atop it. OpenSSL, the cryptographic foundation of the modern web, was maintained by two people with minimal institutional backing until Heartbleed exposed a two-year-old memory safety bug that affected nearly every internet service in existence. Log4j, buried in the dependencies of every major enterprise system on Earth, was maintained by a small team of volunteers. Then, in 2024, a burned-out maintainer of xz utils, a compression library present on billions of computers, was manipulated into granting commit access to someone who embedded a backdoor, nearly becoming the most catastrophic software supply chain attack in history. The pattern was clear: the most critical software in the world was being maintained by people who were drowning.

Zig's design itself was born from this suffering. Kelley had spent years wrestling with C and C++, trying to maintain code that broke in invisible ways, undefined behavior lurking in seemingly harmless operations, hidden allocations that consumed memory without announcement, build systems so fragile they could shatter across different machines. He wanted to create a language that made the treacherous explicit: no hidden control flow, no implicit allocations, explicit error handling. Zig introduced `comptime`, compile-time code execution that let programmers push computation left, eliminating runtime mystery. It made undefined behavior detectable rather than an invisible landmine. It positioned itself as "a better C", C's power and simplicity, but without C's thousand footguns.

The language attracted people who cared about these things: systems programmers tired of fighting their tools, engineers at companies like Uber who needed to cross-compile for different architectures without maintaining a separate build toolchain, financial software builders like TigerBeetle who needed complete control over memory layout and allocation. Bun, a JavaScript runtime, was built in Zig. But none of this adoption happened because Kelley was working full-time on it. It happened because he was working himself to exhaustion at night, because the community believed in the vision enough to contribute, because he had negotiated his contract to own his own work.

What Kelley discovered, living on donations after his leap in June 2018, was that the donation model could actually work, but barely. His community sustained him. Some months were precarious. He lived lean. But he survived, and more importantly, he continued to build. In 2020, two years into full-time donation-funded work, Kelley formalized the arrangement by founding the Zig Software Foundation as a 501(c)(3) nonprofit corporation. Far from a retreat from the donation model, this was a legitimation of it, a way to take the informal community support that had kept him alive and turn it into an institutional structure that could outlast him.

The ZSF eventually attracted real financial backing. Companies like TigerBeetle and Synadia, which depended on Zig working well, pledged $512,000 over two years. But this came only after Kelley had already proven that full-time open-source development on donations was viable, that the community would show up for work that solved real problems in real ways. His gamble had worked, but only because he had been willing to be vulnerable enough to try it.

The emotional weight of this decision haunts the broader conversation. Kelley has spoken openly about the cost of open-source leadership: the grumpiness that comes from triage, the interpersonal tensions with users who don't understand why feature requests go unfunded, the loneliness of being responsible for something that thousands of people depend on without bearing financial responsibility for it. A decade into building Zig, he finds himself confronting the human cost of stewardship rather than celebrating technical milestones, "issue triaging while grumpy," apologizing for the frustration that leaks into technical communities when one person carries too much weight.

This is the paradox Kelley's story lays bare: free software is not actually free. Somebody always pays. For decades, that payment came from maintainers like Steve Marquess at OpenSSL, like the lone person at xz utils, like Kelley before June 2018, in sleep debt, in deferred ambitions, in financial precarity, in the slow erosion of faith in the communities that depend on them. What Kelley did by quitting OkCupid was not invent a new economic model so much as expose the poverty of the old one. He made his dependence visible. He refused to pretend that critical software could be maintained as a hobby, that genius and altruism could substitute for actual resources.

By 2024, the software industry had begun, haltingly, to recognize what Kelley had known in 2018: open-source maintainers need to eat. TigerBeetle's database is critical infrastructure, but it exists only because the company investing in it understood that TigerBeetle depended on Zig working well, and Zig depended on Kelley being paid. Uber's adoption of Zig's cross-compiler meant that engineers across the company could do their work without maintaining their own toolchain. But all of this depended on that moment when Kelley looked at his life and leapt.

The thing holding him back was not fear but permission, permission to believe that he was worth paying for, that the work he loved was worth his survival.

This is what the donation button has meant in open source's long history: the moment when a maintainer stops pretending that unpaid work is sustainable and stands in front of the community and asks, plainly, to be paid. Kelley's donation button was not a cry for help. It was a statement of fact. This is the price of maintaining a language used by Uber, by TigerBeetle, by Bun. This is what it costs to move open source from hobby to stewardship. And this is what software that matters should probably cost.

What happened next

Year	Event
2015	Andrew Kelley begins Zig as a side project while employed at OkCupid
2016	Zig announced publicly; gains early interest in systems programming community
2018	Kelley publishes "I Quit My Cushy Job at OkCupid to Live on Donations to Zig" in June; begins full-time work funded by community donations
2020	Zig Software Foundation established as 501(c)(3) nonprofit to formalize and sustain community support
2021	TigerBeetle, a financial database built in Zig, begins development; Bun JavaScript runtime announced, built in Zig
2023	Uber announces adoption of Zig's cross-compiler for architecture-independent build toolchain
2024	TigerBeetle ships to production after 3.5 years of development; Zig community continues rapid growth
2025	TigerBeetle and Synadia pledge $512,000 to Zig Software Foundation over two years

RAKU 2019

Larry Wall, Perl community ● Active

The Perl 6 project was officially renamed Raku in 2019, representing a complete reimagining of Perl. Raku is dramatically different from Perl 5; it has a robust grammar system, multiple dispatch, and Unicode as a first-class feature. It's beloved by language enthusiasts.

```
say 'Hello, World!'
```

say outputs with newline

Paradigm: Multi-paradigm

Typing: Static

Usage: ●○○○○

Lines: 1

★ Raku took 15 years from announcement (2000) to stable release (2015). Larry Wall's design included grammars as first-class citizens, allowing programmers to define entirely new syntax, effectively building languages within Raku.

A SPOTLIGHT ON RUST

Twenty-One Floors

How a Broken Elevator Led to the Language the White House Recommends

---

In 2006, a twenty-nine-year-old programmer named Graydon Hoare came home to his apartment building in Vancouver and found the elevator was broken, again. Its software had crashed. He lived on the twenty-first floor.

As he climbed the stairs, a thought took shape: it was ridiculous that the people who build software couldn't even make an elevator that runs without crashing. He knew the likely cause. The elevator almost certainly ran on C or C++ code, and those languages give programmers direct control over memory, which means they also give programmers direct access to an entire category of bugs that have plagued systems programming since the 1970s. Buffer overflows. Use-after-free errors. Null pointer dereferences. Microsoft later estimated that roughly seventy percent of all its high-severity security vulnerabilities came from memory safety issues.

That evening, Hoare opened his laptop and started designing a language that would make those bugs structurally impossible, caught at compile time, before the program ever runs. He named it Rust, after the rust fungi, organisms he described as "over-engineered for survival."

For three years he worked alone, building an early prototype in OCaml that barely functioned. Around 2009, he showed it to colleagues at Mozilla. Brendan Eich (who had built JavaScript in ten days back in 1995) saw the potential and championed the project. A small team formed in what they called "the nerd cave." Their goal: build a language fast enough to replace C++ in Firefox's browser engine, but safe enough that an entire category of security vulnerabilities would simply cease to exist.

The breakthrough was the ownership and borrowing system: a set of rules, enforced by the compiler, governing which part of a program "owns" each piece of data and who can access it at any given time. If your code violates the rules, it doesn't compile. Period. The system, designed to prevent memory bugs, simultaneously made data races in concurrent code impossible.

Then came the twist. Hoare himself had grown increasingly distant from the language's direction. The community wanted maximum performance and control; Hoare would have traded some speed for simplicity. By 2013, the gap between his vision and the community's had widened. He stepped down as technical lead and eventually left Mozilla for Apple. In a 2023 blog post, he put it plainly: "The priorities I had while working on the language are broadly not the revealed priorities of the community." The language had outgrown its creator.

The community nearly lost Rust entirely on August 11, 2020, when Mozilla laid off approximately 250 employees, including active Rust team members. For an anxious week, the language's future seemed uncertain. But by February 2021, the Rust Foundation launched with five founding corporate members: Amazon, Google, Huawei, Microsoft, and Mozilla.

The Linux kernel began accepting Rust code in October 2022. Most dramatically, the White House itself weighed in: in February 2024, the Office of the National Cyber Director urged all software developers to adopt memory-safe languages, effectively endorsing Rust's founding premise.

It all started because an elevator crashed and a programmer had to climb twenty-one flights of stairs. He decided that shouldn't happen. Then he spent three years alone, building a tool the world didn't know it needed.

What happened next

Year	Event
2006	Graydon Hoare begins Rust as a personal project at Mozilla
2009	Mozilla officially sponsors Rust; development accelerates
2015	Rust 1.0 released on May 15 with backward compatibility guarantees
2016	Rust named 'most loved' language on Stack Overflow developer survey
2020	Mozilla layoffs briefly threaten Rust's future
2021	Rust Foundation established by Amazon, Google, Huawei, Microsoft, Mozilla
2022	Rust merges into Linux kernel 6.1; first non-C code in the core kernel
2024	White House endorses memory-safe languages, validating Rust's premise

Era 8

The AI Age

2020–2024

The defining paradox of the AI era: the language that dominates artificial intelligence wasn't designed for it. Python, created in 1991 as a general-purpose scripting language, became the de facto standard for AI development not because it's the best tool for the job, but because the entire machine learning ecosystem, TensorFlow, PyTorch, NumPy, scikit-learn, was built on it. By the time anyone noticed, it was too late to switch. The question of the 2020s became: do we fix Python, or replace it?

This period also saw continued maturation and specialization. Rust, barely a decade old, became credible enough for Linux kernel development; the Linux kernel began accepting Rust code in October 2022. In February 2024, the White House itself weighed in: the Office of the National Cyber Director urged all software developers to adopt memory-safe languages, effectively endorsing Rust's founding premise. Gleam brought refined thinking about functional programming and type systems to the Erlang VM. Carbon proposed reimagining C++ for the modern era, acknowledging that even foundational systems languages need reinvention. And Mojo, created by Swift's architect Chris Lattner, aimed to give Python the performance of C for AI workloads.

The era demonstrated that language innovation never stops, and raised a provocative question: what does language design look like when AI can generate code in any language? The languages below represent humanity's latest attempts to talk to machines. They won't be the last.

4 languages in this era

Timeline

- 2020 Paul Chiusano creates Unison
- 2022 Chandler Carruth announces Carbon as C++ successor concept
- 2023 Chris Lattner creates Mojo at Modular
- 2024 Louis Pilfold releases Gleam 1.0

World Context

COVID-19 pandemic; AI boom with ChatGPT and large language models; focus on sustainability and responsible AI

Dominant Paradigms

AI-first thinking; continued emphasis on safety and correctness; functional programming refined; systems languages evolving

UNISON 2020

Paul Chiusano, Rúnar Bjarnason, Arya Irani

Active

Paul Chiusano, Rúnar Bjarnason, and Arya Irani created Unison as a functional language with a content-addressed code model. Instead of files and git, code is stored in a database indexed by content hash, so functions can be serialized and shipped across machines, making distributed computing a first-class capability. Unison reached 1.0 in November 2025 after more than 26,500 commits.

```
main : '{IO, Exception} ()
main _ = printLine "Hello, World!"
```

Type signature with effects

Paradigm: Functional

Typing: Static

Usage: ●○○○○

Lines: 2

★ In Unison, code is stored as a hash-addressed tree, not text files. Renaming a function doesn't break anything because references point to the hash, not the name. There are no merge conflicts for renames.

CARBON 2022

Chandler Carruth, Google

Experimental

Chandler Carruth at Google created Carbon as a 'successor to C++' addressing C++'s decades of accumulated complexity. Carbon is a modern systems language with direct C++ interoperability, enabling incremental migration rather than wholesale rewrites. Memory safety, originally out of scope, was added to the roadmap in 2024. It remains experimental, with a v0.1 MVP expected no earlier than late 2026 and a v1.0 release not anticipated before 2028.

```
fn Main() -> i32 {
  Print("Hello, World!");
  return 0;
}
```

fn Main is the entry point; -> declares the return type

Paradigm: Imperative

Typing: Static

Usage: ●○○○○

Lines: 4

★ Carbon's lead designer Chandler Carruth described it as 'an experiment in C++ interop', Carbon can call C++ code directly, which Rust cannot. Google designed it specifically for the billions of lines of C++ that can't be rewritten.

A SPOTLIGHT ON SWIFT AND MOJO

"From Swift to Mojo: The Architect of Everything"

Chris Lattner's Arc

In 2000, a master's student at the University of Illinois named Chris Lattner began working on a compiler infrastructure project for his thesis. He called it LLVM (Low Level Virtual Machine) and its modular design for code optimization was so elegant that Apple hired him in 2005 to bring it in-house.

LLVM became the compiler infrastructure behind an astonishing number of modern languages. Swift, Rust, Julia, Kotlin (for native), and dozens of others in this book rely on LLVM to turn high-level code into efficient machine instructions. If a modern language compiles code, there's a good chance LLVM is doing the work somewhere in the pipeline.

At Apple, Lattner spent years building Clang, a C/C++ compiler that replaced GCC as the default on macOS and iOS. Then he pitched something more ambitious: a completely new language to replace Objective-C for Apple development. Swift launched at Apple's 2014 developer conference to a standing ovation. Within two years, it had displaced Objective-C as the primary language for iOS, macOS, watchOS, and tvOS development, one of the fastest language transitions in history. Lattner had replaced the dominant language of an entire platform, and he'd done it from inside the company that owned it.

But Lattner wasn't done. He left Apple in 2017 and spent time at Tesla working on the Autopilot team, then at Google Brain working on AI infrastructure. Each stop reinforced the same observation: AI researchers were trapped in what the Julia community calls the "two-language problem." Scientists prototype in Python because it's easy. Then they rewrite everything in C++ for production because Python is too slow. The rewriting wastes months of engineering time and introduces bugs.

In 2022, Lattner founded Modular and announced Mojo, a language designed to eliminate that barrier entirely. Mojo's promise: Python syntax with systems-level performance. It achieves this by compiling Python-compatible code to MLIR (Multi-Level Intermediate Representation), the same compiler infrastructure that powers TensorFlow and PyTorch internally.

One person's career traces the path from "how do we compile code efficiently" to "how do we replace legacy systems languages" to "how do we make AI development practical." Whether Mojo succeeds is an open question. But the ambition is undeniable: to give AI engineers a single language as easy as Python and as fast as C, eliminating the two-language problem that has plagued scientific computing for decades. If anyone has earned the right to attempt it, it's the person who built the compiler infrastructure the rest of the industry runs on.

What happened next

Year	Event
2000	Lattner begins LLVM as his master's thesis at the University of Illinois
2003	LLVM released as open source; grows into dominant compiler infrastructure
2005	Apple hires Lattner; he drives LLVM into production across Apple's toolchain
2007	Lattner architects Clang as a modern replacement for GCC
2010	Lattner begins designing Swift in secret at Apple
2014	Swift announced at WWDC; immediate adoption by iOS developers
2015	Apple open-sources Swift; 60,000+ repository clones in the first week
2017	Lattner leaves Apple; works at Tesla, then Google on AI infrastructure
2022	Lattner co-founds Modular and raises $30 million in seed funding
2023	Mojo announced, promising Python syntax with systems-level performance

MOJO 2023

Chris Lattner, Modular ● Active

Chris Lattner (who created Swift and LLVM) founded Modular and created Mojo as a superset of Python for AI systems. Mojo adds systems programming features and performance optimizations while remaining Python-compatible. With $380 million raised and a $1.6 billion valuation, Modular had attracted 175,000 developers across 50,000 organizations by 2024. A 1.0 release and full compiler open-sourcing are targeted for 2026.

```
def main():
    print("Hello, World!")
```

def main() entry point required; Python-like syntax

Paradigm: Multi-paradigm

Typing: Dynamic/Static

Usage: ●●○○○

Lines: 2

★ Mojo claimed a 68,000x speedup over Python on certain benchmarks by compiling to MLIR, the same compiler infrastructure that powers TensorFlow and PyTorch. Independent benchmarks place compiled Mojo closer to Rust and C · fast, but not the miracle the headlines suggested.

GLEAM 2024

Louis Pilfold ● Active

Louis Pilfold created Gleam as a friendly language for building scalable concurrent systems, running on both the Erlang VM and JavaScript. Gleam combines functional programming with Erlang's reliability. By 2025, it had ranked second in Stack Overflow's 'most admired languages' survey, behind only Rust, a remarkable result for a language that had reached 1.0 just a year earlier.

```
import gleam/io

pub fn main() {
  io.println("Hello, World!")
}
```

Module imports; io.println outputs

Paradigm: Functional

Typing: Static with Inference

Usage: ●○○○○

Lines: 5

★ Gleam compiles to both Erlang bytecode and JavaScript from the same source code. A single Gleam module can run on a server (via the BEAM VM) and in a browser, same types, same guarantees, two runtimes.

A SPOTLIGHT ON COBOL

When the Pandemic Broke the Code That Runs Civilization

On April 4, 2020, with the United States recording over 300,000 confirmed COVID-19 cases and the economy in free fall, New Jersey Governor Phil Murphy held a press conference. Among the urgent topics, hospital capacity, ventilator shortages, stay-at-home orders, he made an unusual request: the state desperately needed volunteers who knew an old programming language.

He called it "Cobalt." He meant COBOL.

The mispronunciation went viral, but the crisis was real. In the previous two weeks, 362,000 New Jerseyans had filed unemployment claims, a volume the state's forty-year-old mainframe systems were never designed to handle. The programs processing those claims were written in COBOL, a language created in 1959 by a committee that included Grace Hopper. The code had been running, continuously, for decades. Nobody had needed to think about it, until everyone did.

New Jersey was not alone. Connecticut, Kansas, and at least a dozen other states reported similar failures. The federal government's systems groaned under identical pressure. An estimated 220 billion lines of COBOL remained in active use worldwide, processing 95 percent of ATM transactions and 80 percent of in-person credit card swipes. The language that everyone thought was dead was, in fact, running the financial infrastructure of civilization.

The problem was not that COBOL was bad. The problem was that the people who knew it were retiring or dying. The average COBOL programmer in 2020 was estimated to be in their fifties or sixties. Universities had stopped teaching it decades ago. When states began calling for volunteers, the people who responded (retirees, hobbyists, a few corporate consultants) were often in the highest COVID risk group, being asked to sit in cramped government offices to debug code written before the claimants filing for help were born.

The irony was brutal. COBOL's design principles, readability, verbosity,

English-like syntax, were specifically intended to make programs maintainable by future generations. Grace Hopper insisted that code should read like English so that anyone could understand it. And COBOL does read like English: MOVE TOTAL-AMOUNT TO PRINT-LINE. ADD 1 TO RECORD-COUNT. The syntax is so clear that a nonprogrammer can follow the logic.

But maintainability requires more than readable code. It requires people who are willing to do the maintaining. By 2020, COBOL had become what the industry calls "legacy", not because it didn't work, but because working on it wasn't glamorous. Young programmers wanted to build apps, not fix mainframes. Computer science departments wanted to teach the future, not the past. The infrastructure survived; the workforce didn't.

In the aftermath, IBM launched free COBOL training programs. Several states initiated modernization projects to rewrite their unemployment systems. A few universities added COBOL back to their curricula. But the deeper lesson reached beyond any single language, to the invisible foundations of modern life: code that works so reliably for so long that everyone forgets it exists, until the moment it can't keep up.

Grace Hopper died in 1992. She never saw her language's most dramatic test. But she would have recognized the problem instantly: the world had done what she spent her career fighting against. It had assumed that because something worked yesterday, it would work forever.

What happened next

Year	Event
1960	COBOL specifications approved; same program runs on both RCA and Univac
1970	COBOL becomes the most widely used programming language in the world
1985	COBOL-85 standardized with structured programming features
2002	Object-oriented COBOL standardized as COBOL 2002
2014	COBOL 2014 published with dynamic tables and improved interoperability
2020	COVID-19 pandemic overwhelms state unemployment mainframes; volunteers called
2023	IBM launches watsonx Code Assistant for Z to modernize COBOL with AI
2025	An estimated 800 billion lines of COBOL remain in production worldwide

Afterword

The first pattern you notice, reading ninety languages in a row, is that none of them had to exist. Every language in this book was born from someone's conviction that the current tools weren't good enough. The cycle is always the same: frustration, insight, design, community, legacy. Some become foundational, C powers the operating systems, Python powers the AI. Some become cautionary tales about overreach; PL/I tried to replace everything and replaced nothing. Some become jokes that accidentally reveal deep truths about computation itself; Brainfuck's eight instructions are enough for Turing completeness, and Malbolge's self-modifying chaos is technically capable of anything any supercomputer can do. But every single one started with someone saying "I can do better."

The second pattern is the constant tension between simplicity and power. The book traces an oscillation that never resolves. FORTRAN was powerful but complex. BASIC was simple but limited. C found the sweet spot for a generation. Python found it for the next. Each era rediscovers that the "right" balance depends entirely on who's programming and what they're trying to build. There is no universal answer, which is why there are ninety languages in this book instead of one.

The third pattern is invisible infrastructure. COBOL runs the banks. FORTRAN runs the climate models. C runs the operating systems. The most important code in civilization is often the oldest and least glamorous, maintained by the smallest and most aging workforce. When the COVID-19 pandemic hit in 2020, state unemployment systems buckled under unprecedented demand. The programs processing those claims had been running, untouched, since before the claimants were born. The language that everyone thought was dead was running the financial infrastructure of civilization.

And now, a question the book can pose but not answer: in 2026, large language models can generate Hello World in any language in this book in milliseconds. Does the program still matter when an AI can write it? The answer is yes, because Hello World was never really about the code. It was about the human. The moment a person types something and a machine responds. That moment of connection, I spoke, and the machine

The Ones That Got Away

Notable languages that didn't receive full entries in this edition.

A-0 (1952): Grace Hopper built the first compiler before anyone believed compilers were possible. A-0 translated mathematical notation into machine code on the UNIVAC I, and Hopper spent years convincing skeptics it actually worked.

JOVIAL (1960): The US Air Force's language for mission-critical embedded systems. JOVIAL ran weapons systems and flight computers for decades, some of it reportedly still flying in B-52 bombers.

CPL (1963): Cambridge and London universities built the Combined Programming Language as an ambitious successor to ALGOL. Too complex to be practical, CPL's ideas flowed through BCPL and B into C, a failure that seeded a dynasty.

Simula 67 (1967): The refined version of SIMULA that fully realized classes, inheritance, and virtual methods. Every object-oriented language descends from what Dahl and Nygaard built in this revision.

BLISS (1970): DEC's systems programming language for the VAX architecture. BLISS compiled to remarkably efficient code and showed that systems languages didn't have to be C, though C won anyway.

Hope (1976): Rod Burstall and Dave MacQueen's functional language that pioneered algebraic data types and pattern matching. Hope was the direct ancestor of ML, which inspired everything after it.

Fortran 77 (1978): The major FORTRAN revision that added character handling and structured IF-THEN-ELSE blocks. Fortran 77 was the workhorse of scientific computing for two decades and some of its code ran climate models for decades afterward.

Modula-2 (1978): Niklaus Wirth's evolution of Pascal added modules and systems programming capability. Widely used in European industry and education, Modula-2 introduced modular organization years before C had anything comparable.

Modula-3 (1989): Refined Modula-2 with garbage collection, exception handling, and object-oriented features. Modula-3 seeded ideas in Java and C#'s designs, but never found its own audience.

FL (1989): John Backus, creator of FORTRAN, spent his final years exploring functional programming. FL was his manifesto against the imperative paradigm he'd helped create, a remarkable act of intellectual reinvention.

ColdFusion (1995): Allaire's tag-based web language powered thousands of corporate intranets in the late 1990s. It made server-side web development accessible to designers, then faded as PHP and frameworks took over.

Rebol (1997): Carl Sassenrath (designer of the Amiga OS) created Rebol around a radical idea: code and data should be the same thing, expressed as human-readable dialects. Beautiful concept, tiny community.

D (2001): Walter Bright's attempt to build 'C++ done right', faster compilation, cleaner syntax, better defaults. D is technically excellent but never overcame C++'s entrenched momentum. A case study in why the best design doesn't always win.

Chef (2002): David Morgan-Mar's esoteric language where programs look like cooking recipes. Variables are ingredients, loops are stirring, and the output is 'served', language design as culinary art.

Pony (2014): A language that guarantees data-race freedom at compile time through a capabilities system. Pony solved one of concurrent programming's hardest problems harmoniously“ but almost nobody noticed.

Solidity (2015): The language of Ethereum smart contracts. Whether you view blockchain as the future of finance or an elaborate speculation, Solidity is the syntax in which billions of dollars of agreements are written.

Wren (2016): Bob Nystrom wrote 'Crafting Interpreters,' one of the best books on building programming languages, and created Wren as its companion language. Clean, tiny, and a masterclass in language implementation.

Acknowledgments

A book about other people's languages depends on those people. Seven creators read drafts of their entries and made the book more accurate.

Entries reviewed and corrected by

Don Chamberlin (SQL)
Martin Richards (BCPL)
Alan Cooper (Visual Basic)
Bjarne Stroustrup (C++)
Simon Peyton Jones (Haskell)
Don Syme (F#)
Ian Buck (CUDA)

Don Chamberlin moved Larry Ellison to the right place in the timeline and reminded me that SQL is the world's most-used query language, not its most-used programming language. Martin Richards read the BCPL essay, called it perfect, then sent two pages of design context that no textbook contains. Alan Cooper read the Visual Basic essay and corrected the irony, explaining that he had spent his career building tools for users and accidentally built the most important tool for programmers. Simon Peyton Jones tightened the Haskell spotlight and then read the full manuscript, suggesting the structural changes that shaped the book's final form: clearer spotlight labels, better-placed sources, and a prominent pointer to the History of Programming Languages proceedings. Ian Buck compiled and ran the CUDA program and removed a line I did not need.

This book had an editor before it had a publisher. Natasha Biagio read every draft, corrected every error she could find, and asked the questions that made every page better. None of this happens without her. Ava Biagio kept me honest about what was actually interesting and what was not. Roberto and Linda Biagio, my parents, taught me that hard work pays off. They were right. And the broader Biagio family carried the project alongside everything else they carry. I am grateful, every day.

Errors that remain are mine.

– *Dale Biagio*

About the Author

Dale Biagio is a software architect with twenty-five years of experience, from early mobile platforms to modern enterprise AI. He has built and led global engineering teams, holds certifications in enterprise architecture, AWS, and Azure AI, and studied Business Analytics, Economics for Managers, and Financial Accounting at Harvard Business School.

He wrote this book because after a quarter century of typing "Hello, World!" in new languages, he wanted to know where those two words actually came from.

He lives in Georgia, USA with his wonderful Wife, fearless Daughter, and one very unprogrammable Dog.

Sources and Further Reading

This book draws on published interviews, official documentation, academic papers, and widely cited industry sources. Key references for the spotlight essays follow.

Assembly Entry

Booth, Kathleen H.V. "Coding for A.R.C." Birkbeck College, University of London, 1947.
Wilkes, Maurice V., Wheeler, David J., and Gill, Stanley. The Preparation of Programs for an Electronic Digital Computer. Addison-Wesley, 1951.
Computer Conservation Society, "Kathleen Booth" biographical note.

Grace Hopper Spotlight

Beyer, Kurt W. Grace Hopper and the Invention of the Information Age. MIT Press, 2009.
Williams, Kathleen Broome. Grace Hopper: Admiral of the Cyber Sea. Naval Institute Press, 2004.
Nanosecond wire demonstration: Hopper's lectures at the National Science Foundation.

John Backus and Fortran Spotlight

Backus, John. "The History of FORTRAN I, II, and III." Annals of the History of Computing, 1978.
Backus, John. "Can Programming Be Liberated from the von Neumann Style?" ACM Turing Award Lecture, 1977.
Computer History Museum. Oral History of John Backus, interviewed by Grady Booch, September 2006.
IBM corporate history: "FORTRAN" and "John Backus" biographical pages.
Haibt, Lois. Biographical profile, Computer History Museum.
The New Stack, "How John Backus' Fortran Beat the Machine Code's Priesthood," 2024.

John McCarthy and LISP Spotlight

McCarthy, John. "Recursive Functions of Symbolic Expressions and Their Computation by Machine, Part I." Communications of the ACM, April 1960.
ACM. "John McCarthy, A.M. Turing Award Laureate."
Two-Bit History. "How Lisp Became God's Own Programming Language," October 2018.
Patel, Nitin. "The Early History of LISP (1956-1959)." Medium.
Graham, Paul. "The Roots of Lisp." paulgraham.com.
Communications of the ACM. "John McCarthy, 1927-2011."
Fogus, Michael. "In the Shadow of John McCarthy." blog.fogus.me, November 2011.
Computer History Museum. "John McCarthy."

COBOL Statistics

Reuters, "Wanted Urgently: People Who Know a 50-Year-Old Computer Language," April 2017.
Phil Murphy press briefing transcript, nj.gov, April 4, 2020.
Gartner/DataPro industry estimates, 1997–2017.

BASIC Spotlight

Kurtz, Thomas. Oral history interview. Rauner Special Collections, Dartmouth College, 2020.
Kemeny, John G. and Kurtz, Thomas E. Back to BASIC. Addison-Wesley, 1985.
TIME, "Fifty Years of BASIC," May 2014.
Dartmouth Faculty of Arts and Sciences, "Remembering Thomas Kurtz," November 2024.

Seymour Papert and Logo Spotlight

Papert, Seymour. Mindstorms: Children, Computers, and Powerful Ideas. Basic Books, 1980.
MIT News. "Professor Emeritus Seymour Papert, pioneer of constructionist learning, dies at 88," August 2016.
MIT Media Lab. "Piaget's Constructivism, Papert's Constructionism: What's the difference?"
The Daily Papert. "About Seymour Papert." dailypapert.com.
EdSurge. "A Glimpse Into the Playful World of Seymour Papert," August 2016.

MIT Media Lab. "Logo History." Logo Foundation.
eLearningInside News. "Seymour Papert, Logo Turtles, and the Origin of Educational Robots."
Voice of America. "Computer Guru's Accident Highlights Vietnam's Traffic Woes," December 2006.

Martin Richards and BCPL Spotlight

Richards, Martin. Personal correspondence with the author, 2026.
Richards, Martin. "BCPL: The Language and Its Compiler." Cambridge University Press, 1980.
Richards, Martin. "The BCPL Cintcode System and Cintpos User Guide." Cambridge, 2019.
Ritchie, Dennis M. "The Development of the C Language." HOPL-II, 1993.
Thompson, Ken. "Users' Reference to B." Bell Labs internal memo, 1972.
Kernighan, Brian W. and Dennis M. Ritchie. The C Programming Language. Prentice Hall, 1978.

Dennis Ritchie and C Spotlight

Ritchie, Dennis M. "The Development of the C Language." Proceedings of the ACM History of Programming Languages Conference (HOPL-II), 1993.
ACM. "Dennis M. Ritchie, A.M. Turing Award Laureate."
Kernighan, Brian W. and Dennis M. Ritchie. The C Programming Language. Prentice Hall, 1978.
Pike, Rob. Tribute to Dennis Ritchie, Google+, October 2011.
Harvard Crimson. "Dennis Ritchie '63, The Man Behind Your Technology," May 2013.
Ritchie, Dennis M. "The Evolution of the Unix Time-sharing System." Bell Labs.
Kernighan, Brian W. "Dennis M. Ritchie." Princeton University archive.

Don Chamberlin, Raymond Boyce, and SQL Spotlight

Chamberlin, Donald D. and Raymond F. Boyce. "SEQUEL: A Structured English Query Language." Proceedings of the 1974 ACM SIGFIDET Workshop, May 1974.
Chamberlin, Donald D. "50 Years of Queries." Communications of the ACM, 2024.
Chamberlin, Donald D. "Early History of SQL." researchgate.net.
Computer History Museum. Oral History of Donald Chamberlin, interviewed by Paul McJones.
History of Information. "Donald Chamberlin & Raymond Boyce Develop SEQUEL (SQL)." historyofinformation.com.
Two-Bit History. "Important Papers: Codd and the Relational Model," December 2017.
DataCamp. "50 Years of SQL with Don Chamberlin, Computer Scientist and Co-Inventor of SQL."
ACM. "Edgar F. Codd, A.M. Turing Award Laureate."

Alan Kay and Xerox PARC Spotlight

Hiltzik, Michael A. Dealers of Lightning: Xerox PARC and the Dawn of the Computer Age. HarperBusiness, 1999.
Kay, Alan. "The Early History of Smalltalk." ACM SIGPLAN Notices 28, no. 3, 1993.
Smith, Douglas K. and Alexander, Robert C. Fumbling the Future: How Xerox Invented, Then Ignored, the First Personal Computer. William Morrow, 1988.
Isaacson, Walter. Steve Jobs. Simon & Schuster, 2011.
Ingalls, Dan. "Design Principles Behind Smalltalk." BYTE Magazine, August 1981.

Bjarne Stroustrup and C++ Spotlight

Stroustrup, Bjarne. "A History of C++: 1979–1991." ACM SIGPLAN Notices 28, no. 3, 1993.
Stroustrup, Bjarne. The Design and Evolution of C++. Addison-Wesley, 1994.
Stroustrup, Bjarne. The C++ Programming Language. Addison-Wesley, 1st edition, 1985.
Computer History Museum. Oral History of Bjarne Stroustrup, interviewed by Paul McJones, February 2015.
Stroustrup, Bjarne. FAQ and Quotes pages at stroustrup.com.

Esoteric Languages Spotlight

Olmstead, Ben. Original Brainfuck specification, aminet.net, 1993.

Joe Armstrong and Erlang Spotlight

Armstrong, Joe. "A History of Erlang." Proceedings of the ACM SIGPLAN Conference on History of Programming Languages (HOPL III), 2007.
Armstrong, Joe. "Making reliable distributed systems in the presence of software errors." PhD thesis, Royal Institute of Technology, Stockholm, 2003.
Alihodzic, Imad. "The Rise & Fall of Erlang at Ericsson AB."
Singh, Ajit. "How WhatsApp Scaled to Billions of Users with Just 50 Engineers." singhajit.com.
The New Stack. "Why Erlang? Joe Armstrong's Legacy of Fault-Tolerant Computing."
The Chip Letter. "Ericsson to WhatsApp: The Story of Erlang."
Erlang Solutions. "Remembering Joe, a Quarter of a Century of Inspiration and Friendship."

Simon Peyton Jones and Haskell Spotlight

Hudak, Paul, John Hughes, Simon Peyton Jones, and Philip Wadler. "A History of Haskell: Being Lazy With Class." Proceedings of the Third ACM SIGPLAN Conference on History of Programming Languages (HOPL III), 2007.
CMU POPL Interview Series. "Simon Peyton Jones." cs.cmu.edu.
Archives of IT. "Simon Peyton Jones Interview." archivesit.org.uk.
Peyton Jones, Simon. "Bio and Pictures." simon.peytonjones.org.
The Royal Society. "Simon Peyton Jones FRS." royalsociety.org.
Yale News. "In memoriam: Paul Hudak, computer scientist and Saybrook College master," April 30, 2015.
Serokell. "Past and Present of Haskell: Interview with Simon Peyton Jones."
Red Gate. "Simon Peyton Jones: Geek of the Week." simple-talk.com.
University of Bath. "Professor Simon Peyton Jones: Oration." bath.ac.uk.
Facebook Engineering. "Fighting spam with Haskell," June 2015.
HaskellWiki. "Haskell in industry."
Computing at School. "Professor Simon Peyton Jones receives OBE," May 2022.

Guido van Rossum and Python Spotlight

Computer History Museum. Oral History of Guido van Rossum, interviewed by Hansen Hsu, February 2018.
Van Rossum, Guido. "The History of Python" blog at python-history.blogspot.com.
Peters, Tim. "The Zen of Python." PEP 20, Python Enhancement Proposals, 2004.
Van Rossum, Guido. BDFL resignation statement, python.org mailing list, July 12, 2018.
Lex Fridman Podcast #341, interview with Guido van Rossum, November 2022.
Talk Python To Me, Episode 100, interview with Guido van Rossum, January 2017.

Larry Wall and Perl Spotlight

Wall, Larry. "State of the Onion" addresses, 1997–2015. Available at perl.com.
Wall, Larry, Christiansen, Tom, and Orwant, Jon. Programming Perl. O'Reilly, 3rd edition, 2000.
Big Think, "Perl Founder Larry Wall Explains His Po-Mo Creation," 2011.
The New Stack, "Larry Wall's Quest for a 100-Year Programming Language," 2016.
BioPerl project, "How Perl Saved the Human Genome Project," bioperl.org.

Yukihiro Matsumoto and Ruby Spotlight

Two-Bit History. "The Ruby Story," November 2017.
Venners, Bill. "The Philosophy of Ruby: A Conversation with Yukihiro Matsumoto." Artima, 2003.
Ruby Language. "About Ruby." ruby-lang.org.
Auth0. "A Brief History of Ruby."
Karosic, Laina. "7 Facts About Ruby and its Creator, Yukihiro 'Matz' Matsumoto." Medium.
SitePoint. "The History of Ruby."

Alan Cooper and Visual Basic Spotlight

Computer History Museum. "2017 CHM Fellow Alan Cooper: Father of Visual Basic."
Cooper, Alan. "Plan to Throw One Away: A Story about Ruby." Medium.
Forest Moon Productions. "The Birth of Visual Basic." forestmoon.com.
YourStory. "The untold story of Alan Cooper, the father of Visual Basic," June 2017.
Computer History Museum. "Alan Cooper."

JavaScript Spotlight

Severance, Charles. "JavaScript: Designing a Language in 10 Days." IEEE Computer, Feb 2012.
Eich, Brendan. Public talks and blog posts at brendaneich.com.

James Gosling and Java Spotlight

Computer History Museum. Oral History of James Gosling, part 1 of 2, interviewed by Hansen Hsu, September 2019.
Computer History Museum. Oral History of James Gosling, part 2 of 2, interviewed by Hansen Hsu, September 2019.
Naughton, Patrick. "Java: The Inside Story." tech-insider.org, July 1995.
"A Brief History of the Green Project." Sun Microsystems internal history, tech-insider.org, May 1998.
Sun Microsystems, Inc. v. Microsoft Corp., 188 F.3d 1115 (9th Cir. 1999).
Gosling, James. Blog post on resignation from Oracle, nighthacks.com, April 2, 2010.
eWeek, "Java Creator James Gosling: Why I Quit Oracle," September 2010.
eWeek, "Java Creator Gosling: Oracle's Android Lawsuit Is No Surprise," August 2010.
The New Stack, "Java's James Gosling on Fame, Freedom, Failure Modes and Fun," 2022.
Building Nubank, "A Mind-Blowing Conversation with James Gosling, Java's Father," building.nubank.com.
The New Stack, "Java at 30: The Genius Behind the Code That Changed Tech," 2025.

Anders Hejlsberg and C# Spotlight

Hejlsberg, Anders. dotJS 2018 conference talk. Genesis of C# and .NET.
Hejlsberg, Anders. "The C# Design Process." Artima interview with Bruce Eckel, 2003.
Hejlsberg, Anders. Aarthi & Sriram Show, Episode 34, 2023.
Biancuzzi and Warden. Masterminds of Programming. O'Reilly, 2010. Chapter 13: C#.
Sun Microsystems v. Microsoft Corp., 188 F.3d 1115 (9th Cir. 1999).
Microsoft–Sun settlement announcement, news.microsoft.com, January 2001.
The Register, "Happy Birthday: .NET turns 20," February 2022.

Don Syme and F# Spotlight

Syme, Don. "The Early History of F#." Proceedings of the ACM on Programming Languages, Vol. 4, No. HOPL, Article 75, June 2020.
Kennedy, Andrew and Don Syme. "Design and Implementation of Generics for the .NET Common Language Runtime." PLDI, 2001.
Syme, Don et al. "Threads, Events, and Tasks." Submitted to ICFP, 2011.
Syme, Don. "Introducing F# Asynchronous Workflows." Blog post, October 10, 2007.
Hoban, Luke. "The first TypeScript demo." hackernoon.com, 2018.
Torgersen, Mads. "Why you should use F#." Microsoft Channel 9, Build 2017.
Haller, Philipp and Martin Odersky. "Scala Actors: Unifying thread-based and event-based programming." Theoretical Computer Science, 2009.
FSSF Contributors. "F# Testimonials." fsharp.org/testimonials.
Wecker, Dave. "LIQUi|>: A quantum simulator for F#." Microsoft Quantum Computing.
Abraham, Isaac. "SAFE Stack." safe-stack.github.io.

Jeffrey Snover and PowerShell Spotlight

The Register. "Jeffrey Snover claims Microsoft demoted him for PowerShell," May 2022.
CoRecursive Podcast. "Building PowerShell with Jeffrey Snover."
Snover, Jeffrey. "The Monad Manifesto." jsnover.com, October 2011.
Microsoft Learn. "The Monad Manifesto."
The Register. "PowerShell architect retires after decades at the prompt," January 2026.
Cook, John D. "Comparing the Unix and PowerShell pipelines," June 2009.
Microsoft Learn. "About Pipelines."

Rich Hickey and Clojure Spotlight

Hickey, Rich. "A History of Clojure." Proceedings of the ACM on Programming Languages (HOPL IV), 2020.
Hickey, Rich. "Simple Made Easy." Presented at Strange Loop, October 2011. InfoQ.

Clojure. "History." clojure.org.
Higher-Order Blog. "Please Help Funding Clojure," December 2009.
Linux Journal. "Economy Size Geek: Interview with Rich Hickey."
Changelog. "Rich Hickey's Greatest Hits."
Building Nubank. "Clojure's Journey at Nubank: A Look into the Future."

CUDA and NVIDIA Spotlight

Buck, Ian et al. "Brook for GPUs." ACM SIGGRAPH 2004 Papers.
Krizhevsky, Sutskever, Hinton. "ImageNet Classification with Deep Convolutional Neural Networks." NeurIPS, 2012.
NVIDIA corporate history and SEC filings.

Graydon Hoare and Rust Spotlight

MIT Technology Review, "How Rust Went from a Side Project to the World's Most-Loved Programming Language," February 2023.
The New Stack, "Graydon Hoare Remembers the Early Days of Rust," 2023.
Hoare, Graydon. Personal blog at graydon2.dreamwidth.org.
White House ONCD, "Back to the Building Blocks," February 2024.

Chris Lattner Spotlight

Lattner, Chris and Adve, Vikram. "LLVM: A Compilation Framework for Lifelong Program Analysis & Transformation." CGO 2004.
Apple WWDC 2014 Keynote. Swift announcement and developer documentation.
Lattner, Chris. Personal blog and technical posts at nondot.org/sabre.
Modular AI blog and Mojo documentation at modular.com.

Andrew Kelley and Zig Spotlight

Kelley, Andrew. "I Quit My Cushy Job at OkCupid to Live on Donations to Zig." andrewkelley.me, June 2018.
Zig Software Foundation. ziglang.org/zsf.
TigerBeetle. "Synadia and TigerBeetle Pledge $512,000 to the Zig Software Foundation," October 2025.
Uber Engineering. "Hermetic CC Toolchain: Uber's Zig Integration for C/C++ Build."
Cloud Odyssey. "The XZ Backdoor: A Near-Catastrophic Supply Chain Attack and the Fragility of Open Source."
CISA. "Lessons from XZ Utils: Achieving a More Sustainable Open Source Ecosystem."
ProPublica Nonprofit Explorer. "Zig Software Foundation Inc."

General

Code samples verified against Rosetta Code and the Hello World Collection.
Language dates cross-referenced with official documentation and release notes.

Sources and Further Reading (continued)

Yukihiro Matsumoto and Ruby Spotlight (additional)

Wikipedia. "Ruby (programming language)."
Wikipedia. "David Heinemeier Hansson."
Wikipedia. "Ruby on Rails."
Wiktionary. "MINASWAN."
Ruby Language. "Ruby Releases." ruby-lang.org.

Alan Cooper and Visual Basic Spotlight (additional)

Wikipedia. "Alan Cooper (software designer)."
Alan Cooper, personal correspondence with the author, 2026.

Rich Hickey and Clojure Spotlight (additional)

Data Sorcery with Clojure. "Funding Open Source Projects," December 2009.

Andrew Kelley and Zig Spotlight (additional)

Zig Programming Language. "Overview." ziglang.org.
Sun, Ruoyu. "Using Zig As Cross Platform C Toolchain," February 2022.

Sources and Further Reading (continued)

Seymour Papert and Logo Spotlight (additional)

Wikipedia. "Seymour Papert."

Dennis Ritchie and C Spotlight (additional)

Wikipedia. "The C Programming Language."
Wikipedia. "Dennis Ritchie."

Don Chamberlin, Raymond Boyce, and SQL Spotlight (additional)

History of Information. "Donald Chamberlin & Raymond Boyce Develop SEQUEL (SQL)." historyofinformation.com.
Wikipedia. "Raymond F. Boyce."
Chamberlin, Donald D. "50 Years of Queries." Communications of the ACM, 2024.
Your Tech Story. "The Story of Two Genius Scientists Behind SQL," August 2019.
Wikipedia. "Donald D. Chamberlin."
DataCamp. "50 Years of SQL with Don Chamberlin, Computer Scientist and Co-Inventor of SQL."
Wikipedia. "Edgar F. Codd."
Two-Bit History. "Important Papers: Codd and the Relational Model," December 2017.
IBM. "Edgar F. Codd." ibm.com/history.
IBM. "The Relational Database." ibm.com/history.
Wikipedia. "Boyce-Codd Normal Form."
ACM. "Edgar F. Codd, A.M. Turing Award Laureate."
Cockroach Labs. "The Codd Father."
Chamberlin, Donald D. "Early History of SQL." researchgate.net.
IBM Research. "SEQUEL: A Structured English Query Language."

Joe Armstrong and Erlang Spotlight (additional)

Wikipedia. "Joe Armstrong (programmer)."

Alan Cooper and Visual Basic Spotlight (additional)

Wikipedia. "Alan Cooper (software designer)."
Cooper, Alan. "Plan to Throw One Away: A Story about Ruby." Medium.
YourStory. "The untold story of Alan Cooper, the father of Visual Basic," June 2017.

Yukihiro Matsumoto and Ruby Spotlight (additional)

Wikipedia. "Yukihiro Matsumoto."

Jeffrey Snover and PowerShell Spotlight (additional)

Wikipedia. "Jeffrey Snover."

Rich Hickey and Clojure Spotlight (additional)

Wikipedia. "Rich Hickey."
Changelog. "Rich Hickey's Greatest Hits."

HOPL Conference Proceedings

The four History of Programming Languages (HOPL) conference proceedings volumes are indispensable primary sources for any reader who wants to go deeper into the academic record of programming language design. They are cited here in full for the book's bibliography.

HOPL I (1978). History of Programming Languages. Edited by Richard L. Wexelblat. Proceedings of the ACM SIGPLAN History of Programming Languages Conference, Los Angeles, CA, June 1 to 3, 1978. Published by Academic Press / ACM, 1981 (ACM Monograph Series). ISBN 0-12-745040-8.

HOPL II (1993). The Second ACM SIGPLAN Conference on History of Programming Languages. Conference Chair John A. N. Lee. Program Chair Jean E. Sammet. Cambridge, MA, April 20 to 23, 1993. ACM Proceedings DOI 10.1145/154766. https://dl.acm.org/doi/proceedings/10.1145/154766. ISBN 978-0-89791-570-0. Book edition: History of Programming Languages, edited by Thomas J. Bergin and Richard G. Gibson. Addison-Wesley, 1996.

HOPL III (2007). Proceedings of the Third ACM SIGPLAN Conference on History of Programming Languages. Conference Co-Chairs Barbara Ryder and Brent Hailpern. San Diego, CA, June 9 to 10, 2007. ACM Proceedings DOI 10.1145/1238844. https://dl.acm.org/doi/proceedings/10.1145/1238844. ISBN 978-1-59593-766-7.

HOPL IV (2021). Proceedings of the ACM on Programming Languages, Volume 4, Issue HOPL, June 2020. Co-Chairs Guy L. Steele Jr. and Richard P. Gabriel. Conference held virtually June 20 to 22, 2021 (postponed from 2020 due to COVID-19). Open Access. https://dl.acm.org/toc/pacmpl/2020/4/HOPL.

In addition, Simon Peyton Jones and co-authors' "A History of Haskell: Being Lazy with Class" appears in HOPL III (2007). DOI: 10.1145/1238844.1238856. https://dl.acm.org/doi/10.1145/1238844.1238856. This is the definitive citation for the Haskell spotlight.

Typeset in Lora, Liberation Sans, and Noto Sans Mono.

Language entries researched and verified against official documentation, Rosetta Code, and the Hello World Collection.

90 languages · 76 years · One line of code

helloworldthebook.com

www.ingramcontent.com/pod-product-compliance
Lightning Source LLC
LaVergne TN
LVHW081316110826
845149LV00006B/1525

* 9 7 9 8 9 9 5 0 7 2 1 0 2 *